Trademark Without a Lawyer

Thomas Heinz Meitinger

Trademark Without a Lawyer

Your Guide to Securing Valuable Trademarks

Springer

Thomas Heinz Meitinger
Meitinger Patentanwalts GmbH
Munich, Germany

ISBN 978-3-662-73126-0 ISBN 978-3-662-73127-7 (eBook)
https://doi.org/10.1007/978-3-662-73127-7

This book is a translation of the original German edition "Ohne Anwalt zur Marke" by Thomas Heinz Meitinger, published by Springer-Verlag GmbH, DE in 2021. The translation was done with the help of an artificial intelligence machine translation tool. A subsequent human revision was done primarily in terms of content, so that the book will read stylistically differently from a conventional translation. Springer Nature works continuously to further the development of tools for the production of books and on the related technologies to support the authors.

This Springer imprint is published by the registered company Springer-Verlag GmbH, DE, part of Springer Nature.
The registered company address is: Heidelberger Platz 3, 14197 Berlin, Germany

Preface

Registering a trademark with a patent office is not difficult. However, this does not automatically result in a valuable trademark. Acquiring a valuable trademark requires a systematic approach, which typically does not arise intuitively. On the contrary, inexperienced applicants can easily take the wrong path when creating a trademark. In such cases, acquiring a valuable trademark is almost impossible.

A trademark application also carries the risk that, after a few years—when the break-even point has long been surpassed—a holder of an older trademark right may appear and prohibit further use of the mark. For the owner of the more recent trademark, this would be a disaster.

This technical book fulfills two main purposes. First, it provides guidance on how to maintain valuable trademarks. In addition, it aims to equip trademark applicants with the tools necessary to avoid the dreaded scenario of being prohibited from using their own trademark.

This technical book provides comprehensive know-how on the registration and administration of a trademark, from finding a brand name, determining the goods and services, to choosing the type of trademark. It explains at which patent office a trademark can be registered and how to conduct a trademark search. Trademark monitoring and enforcement are further focal points. In addition, the legal options for defending against conflicting trademarks and maintaining one's own trademark rights are explained.

Munich
in June 2021

Patent Attorney Dr. Thomas Heinz
Meitinger

Laws

BGB	German Civil Code in the version promulgated on January 2, 2002 (Federal Law Gazette I p. 42, 2909; 2003 I p. 738), last amended by Article 10 of the Act of March 30, 2021 (Federal Law Gazette I p. 607).
GVG	Courts Constitution Act in the version promulgated on May 9, 1975 (Federal Law Gazette I p. 1077), last amended by Article 4 of the Act of March 9, 2021 (Federal Law Gazette I p. 327).
Markengesetz	of October 25, 1994 (Federal Law Gazette I p. 3082; 1995 I p. 156; 1996 I p. 682), last amended by Article 1 of the Act of December 11, 2018 (Federal Law Gazette I p. 2357).
PVÜ	Paris Convention for the Protection of Industrial Property of March 20, 1883, revised at BRUSSELS on December 14, 1900, at WASHINGTON on June 2, 1911, at THE HAGUE on November 6, 1925, at LONDON on June 2, 1934, at LISBON on October 31, 1958, and at STOCKHOLM on July 14, 1967, and amended on October 2, 1979.
RVG	Lawyers' Remuneration Act of May 5, 2004 (Federal Law Gazette I p. 718, 788), last amended by Article 3 of the Act of June 2, 2021 (Federal Law Gazette I p. 1278).
UMV	Regulation (EU) 2017/1001 of the European Parliament and of the Council of June 14, 2017 on the European Union trade mark.
ZPO	Code of Civil Procedure in the version promulgated on December 5, 2005 (Federal Law Gazette I p. 3202; 2006 I p. 431; 2007 I p. 1781), last amended by Article 8 of the Act of December 22, 2020 (Federal Law Gazette I p. 3320).

Contents

About the Author

Dr. Thomas Heinz Meitinger is a German and European patent attorney. He is Managing Director of Meitinger Patentanwalts GmbH. Meitinger Patentanwalts GmbH is a medium-sized patent law firm based in Munich. After studying electrical engineering in Karlsruhe, he initially worked as a development engineer. Subsequent positions included roles as production manager and technical director in medium-sized companies. Dr. Meitinger regularly publishes scientific articles, writes specialist books on intellectual property law, and gives lectures on topics related to patent and trademark law. Dr. Meitinger holds the degrees Dipl.-Ing. (Univ.) and Dipl.-Wirtsch.-Ing. (FH). In addition, he holds the following master's degrees: LL.M., LL.M., MBA, MBA, M.A., and M.Sc.

Abbreviations

BGH	Federal Court of Justice
BPatG	Federal Patent Court
DPMA	German Patent and Trademark Office
EU	European Union
EuG	General Court of the European Union
EuGH	European Court of Justice
EUIPO	European Union Intellectual Property Office
USPTO	United States Patent and Trademark Office
WIPO	World Intellectual Property Organization

List of Figures

List of Tables

Introduction 1

Important aspects of trademark law are discussed to provide guidance for acquiring a valuable trademark.

1.1 Register Your Trademark!

A trademark should definitely be registered when you intend to start a serious business activity. A trademark is an important starting point for taking action against competitors or for defending your own business activities. The costs of registering a trademark are manageable. For start-ups and entrepreneurs launching a new product or service, registering a trademark is an absolute must.

Example

Best Software GmbH uses the trademark "Pasquale" for its financial accounting software product for small and medium-sized enterprises. They forgot to register the trademark with the patent office. Bad GmbH registers the trademark "Pasquale" for itself and uses it for its financial accounting software. In addition, Bad Software GmbH prohibits Best Software GmbH from continuing to use the "Pasquale" trademark. Best Software GmbH must now find another trademark to market its software product.◄

If you do not register your own trademark with the patent office before offering and selling your products under that trademark, you risk being forced to cease your business

T. H. Meitinger, *Trademark Without a Lawyer*,
https://doi.org/10.1007/978-3-662-73127-7_1

activities. A third party can register and use the trademark for themselves and prohibit the previous user from further use.

1.2 Imaginative Trademarks are Strong Trademarks

Applicants regularly look for a trademark that describes the product or service to be marked. In practice, applicants often strive for a descriptive trademark, believing this will result in a strong and valuable brand. This is a misconception. A descriptive trademark will never become a strong trademark. In fact, it is questionable whether it will even be registered.

It takes a certain effort and determination for the applicant to detach themselves from what is to be designated and to find a trademark that has nothing to do with the goods to be marked. Unfortunately, applicants often believe that only a trademark that describes the product is valuable. Therefore, they aim to protect a generic term as a trademark. However, this is a misguided approach that only becomes apparent over time. Eventually, it becomes clear that a descriptive trademark is not a valuable asset because it does not create brand recognition. A trademark must be imaginative to become valuable.

Table 1.1 compares examples of weak and strong trademarks for a product or service. The comparison helps to develop a sense of what constitutes a strong or weak trademark. If the trademark is closely related to the product to be marked, it is weak. A weak trademark is therefore usually descriptive. Through modifications and additions, such a sign may possibly become registrable. However, it will always remain a trademark with only a small scope of protection. In contrast, a strong trademark is detached from the goods and services to be marked. Most often, a strong trademark is a coined word with no reference to the goods or services.

The strong trademarks in Table 1.1 may not be particularly melodious and, for this reason, may not be suitable as trademarks. They may still lack the "finishing touch." At the very least, they point in the right direction. It is important to understand that, when choosing a name, at least a partial detachment from the goods and services to be marked is necessary in order to arrive at a valuable trademark.

1.3 Please Don't!

In practice, it often happens that a trademark is sought that describes the goods and services to be produced or offered. For example, a trademark "BestSoft" is chosen for software. Then, some graphic elements are added to make the trademark appear more attractive and creative. In the end, you have a word/picture mark with a text component that is descriptive of the goods and services, plus a few graphic elements. In the worst

Table 1.1 Comparison of weak and strong trademarks

Product or Service	Weak Trademarks	Strong Trademark
Cakes and Pastries	"tasty-and-sweet", "always-good", "fresh-from-the-baker", "always-tasty"	Lerus
Restaurant Service	"Pizza-and-more", "Italian-Pizza", "from-the-Italian", "the-real-Italian", "my-Italy"	Bella Mama
Ice Cream Parlor Service	"Ice-Creme", "Italian Icecreme", "Italian Ice", "Bella Italia"	Lombisto
Beverage	"super-smooth", "icy-good", "Alcohol at its finest"	Nadama
Online Wine Retail	"WineService", "WineDelivery-fast", "Wine24/7", "Online-Wine", "WineParadise", "best wines"	Amor
Cosmetics	"myBeauty", "Young-and-Beautiful", "younger + Beauty", "Colors of Beauty", "perfect eyes", "perfect lips"	Yahuda
Travel Agency Service	"beautiful travel", "world travel", "pathfinder to the world", "distant worlds", "far and fast", "travel comfort", "wanderlust", "education through travel", "travel, fun and friends", "travel adventure", "thousand travel worlds", "fascination travel"	Kailu
Auto Repair Shop Service	"Instant-Car-Repair", "Fix-Car", "Fix-Checkup", "Car24/7", "cheap-car-repair", "fast-and-cheap", "Fix-Repair", "Quick-Car-Checkup", "24-h-Service"	Thor
Tire Retail	"24-h-Tire-Retail", "Tires24/7", "wheels-and-more", "TireExpert", "Tires-only-from-the-professional", "Tires-cheap-to-buy"	Crab
Mechanical Feed Device	"120 % Quality", "Feed-from-the-expert", "Correct-and-clean"	Zorx
Bicycle	"myVelo", "Velo24/7", "myBicycle", "myBestBicycle", "Bicycle!!!"	MyMorningStar
Woodworking Machine	"WoodGood", "WoodFast", "SuppperHoolz", "Wood-and-more", "WoodMachine"	Marax
Chemical Intermediate	"Chem24/7", "chemistry-and-more", "Top-Chem"	Ceracux
Turnkey Prefabricated House	"turnkey-and-good", "turnkey-and-done", "Carefree-and-happy-for-a-lifetime", "carefree-living-in-prefab-house", "1-2-3-prefab-house", "good-fast-prefab-house", "key-to-happiness"	MuXXXL

(continued)

Table 1.1 (continued)

Product or Service	Weak Trademarks	Strong Trademark
Roofer Service	"Best-Roofs", "Roofing-at-its-best", "Top-Roof", "Roof2000", "24-7-Roof", "Roof24/7", "Quick-Roof-Repair", "Super-Roof"	Sorom
Plumber Service	"Pipe-clear", "Pipe-fix", "Pipe-and-good", "24/7-Pipes", "Pipe-24-7", "24h-Pipe-clear"	Rex
Architect Service	"House-to-happiness", "Sky-House", "Best-House", "Safety-and-more", "myHouse", "House-clear", "House-ok", "OK-House", "Houses-and-more"	YoorBoon
Textile Product	"myTextile-Product", "Future-of-textile", "Time for Textiles", "Textiles-Quality", "TextileFuture", "TextileGuild", "Textiles-for-the-future"	Mariba
Online Clothing Retail	"Shirts-and-more", "Shirt2000", "Shirt3000", "Shirts-and-socks", "wearables", "best-wearables", "Clothes-just-for-me", "Clothes-are-fun", "Clothes-for-the-family", "Online-Clothing", "Clothing-Express", "Clothing.de", "Clothing.eu", "Clothing.net"	Softex
Software	"Best-Software", "Software-for-you", "Application-for-SMEs", "Software.de", "123Software", "Accounting-in-123"	Pasqale
Web Designer Service	"Webdesigner-only-for-you", "myWebdesign-with-you", "House-for-Webdesign", "Webdesign-Room", "webdesign-for-the-future", "SuperHero Webdesign"	Alhambra
Computer Retail	"Computer-for-you", "mybestComputer", "Computer-for-us-all", "PC-and-so-much-more"	Zeus
Online Toy Retail	"Toys-for-us", "Toys-and-much-more", "Toys-online", "Toys-and-us", "Click-and-Toys", "Clicks-and-toys", "Fix-Toy", "Fast, faster, toys", "all-toys"	Grip

case, the graphic elements are also descriptive of the goods and services for which the trademark is to be used, or else they are simply common embellishments.

At this point, everything has gone wrong! A word/picture mark has been chosen that, by its nature, has only a small scope of protection. As a rule, word/picture marks cannot achieve a broad scope of protection. The text component is largely descriptive, which also keeps the scope of protection small. The graphic elements further reduce the scope of protection. From a trademark law perspective, this is a trademark with a small scope of protection, and thus a trademark with little economic value.

Fig. 1.1 Steps to a valuable trademark

Steps to a valuable brand
• Name creation (imaginative brand) • Overcoming obstacles to registration (distinctiveness, need to keep names free) • No likelihood of confusion with older rights

1.4 Steps to a Valuable Trademark

The most important steps on the way to a valuable trademark are explained (see Fig. 1.1).

1.4.1 Finding a Name

A strong trademark is an imaginative trademark. Descriptive terms are often not eligible for registration. If they are registered, they only provide a small scope of protection. A unique selling proposition cannot be established. Imaginative trademarks, on the other hand, offer broad protection and can be imbued with emotions to convey advertising messages. Imaginative trademarks are therefore inherently suiTable to become strong trademarks.

Example

A trademark like "affordable software" or "software available 247" will never provide broad protection. Purely imaginative names such as "Pasquale," "Manos," or "Alhambra" are better choices.◄

A prominent example from the software industry is Apple Corporation. Through imaginative naming, Apple was able to build a strong trademark with broad protection.

1.4.2 Registrability

A sign is only entered in a trademark register if it overcomes the absolute grounds for refusal.[1] In particular, a sign must be distinctive. A trademark is distinctive if the relevant public perceives it as an indication of origin. In addition, a trademark must not be descriptive of the goods and services for which it is registered (no infringement of the need to keep terms free for general use).

[1] Section 8 of the Trademark Act or Article 7 of the EU Trademark Regulation.

1.4.3 Infringement of Prior Rights

A trademark must not be likely to be confused with an earlier right. Otherwise, the trademark can be removed from the register or will not be registered in the first place.

2 Basics of Trademark Law

The fundamental aspects of trademark law are explained. This facilitates an introduction to the subject area.

2.1 What is a Trademark?

A trademark serves to indicate the origin of a product or service. Only if the product or service has a name that distinguishes it from those of other products and services can it be advertised.

Example

Don Alfredo runs an ice cream parlor. Don Alfredo wants to advertise in a local newspaper. To do this, he places an ad in which he promotes ice cream at the "ice cream parlor." The newspaper readers get a craving for ice cream and go to the nearest ice cream parlor, which is not always Don Alfredo's.◄

The advertising campaign failed. If Don Alfredo had given his service and products a distinctive name, the outcome would have been different.

Example

Don Alfredo names his ice cream parlor "Valentino's" and registers the trademark "Valentino's" for goods in class 30 (ice cream, ice cream products, frozen yogurt,

T. H. Meitinger, *Trademark Without a Lawyer*,
https://doi.org/10.1007/978-3-662-73127-7_2

sorbets, coffee, tea, and cocoa) and services in class 43 (catering services) with the German Patent Office.◀

With a trademark, the goods and services of one company can be distinguished from those of another, thus enabling advertising. A trademark can in particular be one or more words including personal names, images, letters, numbers, sounds, three-dimensional designs including the shape of a product or its packaging, as well as other presentations including colors and color combinations.[1] A trademark can therefore be a word, a slogan, or a logo. By registering a trademark, the trademark owner acquires an exclusive right to the registered trademark.[2] Any third party is prohibited from using an identical trademark for the same goods and services or a trademark that leads to a likelihood of confusion in the course of trade.[3] In addition, there are well-known trademarks in the country that enjoy special protection.[4] Examples of well-known trademarks in Germany include "Coca-Cola," "Apple," or "Mercedes-Benz."

2.2 Registered Trademark

A registered trademark is a trademark filed with a patent office that, after examination, has been entered into the trademark register. A registered trademark provides protection for the territory of a country or for a region, i.e., for the territories of a group of countries. For example, a trademark can be acquired for the territory of Germany or for the territories of the EU member states (Union trademark).

2.3 Trademark by Use

Trademark rights can alternatively arise through the use of a trademark. However, the use must be so intensive that the trademark achieves market recognition.[5] Market recognition of a trademark exists if 20–25% of the relevant public are familiar with the trademark.[6] "Familiar" in this context means that the trademark is associated with the owner. Market recognition is to be distinguished from acquired distinctiveness. With market recognition, trademark rights arise for a sign that has the qualities to be eligible for entry in a trademark register in principle.

[1] Section 3(1) Trademark Act.

[2] Section 14(1) Trademark Act.

[3] Section 14(2) Nos. 1 and 2 Trademark Act.

[4] Section 14(2) No. 3 Trademark Act.

[5] Section 4 No. 2 Trademark Act.

[6] Henning, Piper, Industrial Property Protection and Copyright, 1996, pp. 429–433; Noelle-Neumann, Industrial Property Protection and Copyright, 1966, pp. 70–81.

Market recognition versus market penetration	
Trademark is fundamentally registrable	Trademark is fundamentally not registrable
Market recognition leads to a trademark acquired through use	Market penetration overcomes lack of registrability
20-25% awareness in the relevant trade circles	Over 50% awareness in the relevant trade circles
Regional recognition is sufficient	Nationwide recognition required

Fig. 2.1 Market recognition versus acquired distinctiveness

Acquired distinctiveness is required for trademark rights to arise for a trademark that is not eligible for registration, i.e., one that would not overcome the absolute grounds for refusal upon application. The two most important absolute grounds for refusal are lack of distinctiveness and infringement of the need to keep certain signs free. Acquired distinctiveness requires a higher level of recognition compared to market recognition (see Fig. 2.1).

2.4 Company Signs

Example

The company Pasquale GmbH sells software. The designation "Pasquale" represents the company sign.◄

A company sign is a business designation.[7] A company sign is used in business transactions as a company name or as a special designation of a business operation or company.[8] The prerequisite for the protection of a company sign is that the relevant public perceives the company sign as the designation of a company. If this is not the case, trademark rights in the company sign only arise through market recognition within the relevant public.

[7] Section 5(1) Trademark Act.

[8] Section 5(2) sentence 1 Trademark Act.

2.5 Overview of Trademark Types

The most common type of trademark is the word mark. This is followed by figurative marks, such as a logo, and combined word/figurative marks. Word/figurative marks contain both a text and an image component.

2.5.1 Word Mark

The great popularity of a word mark can at least partly be attributed to its broad scope of protection. The scope of protection of a word mark covers all other later word marks with the same distinctive text, regardless of the font, font size, or other color or graphic design of the text. The scope of protection also extends to word/figurative marks with the same distinctive text component, unless the image component of the word/figurative mark is dominant and the text component is small compared to the image component. A word mark should be the preferred form of trademark.

2.5.2 Word/Figurative Mark

In a word/figurative mark, a text component is registered together with a graphic component. The graphic component can, for example, be a logo. A word/figurative mark is registered if the graphic component is important and should be protected, or if the text component alone is not eligible for protection.

If a word/figurative mark is being considered, it should be determined which component of the mark is the distinctive one. If the text component is the distinctive part, a word mark is recommended. If there is a distinctive image component, a figurative mark should be registered. As a rule, it makes sense to register the significant part of a word/figurative mark as a trademark. The other component only reduces the scope of protection. In case law, the text component of a word/figurative mark is generally regarded as the distinctive component.

Therefore, the image component is not, or hardly, taken into account when determining the scope of protection. The image component of a word/figurative mark therefore generally does not result in protection for the graphic element of the word/figurative mark.

The word/figurative mark should be registered in black and white. This allows the scope of protection to be extended to all color designs. However, special color designs that are not common may still fall outside the scope of protection of a black-and-white word/figurative mark. Very unusual colors of the word/figurative mark can lead to a separate scope of protection, and such a mark with a special color design no longer falls within the scope of protection of the black-and-white word/figurative mark.

	Database	DB	?	DE
111	Registration number	RN	?	1144921
210	Old file number	AKZ	?	C38547
540	Brand representation	MD	?	Coca-Cola
550	Brand form	MF	?	Word and design mark
551	Trademark category	MK	?	Individual trademark
350	Seniority	SEN SENVT	? ?	Union trademark file number: 2107118 Date of publication of seniority: 08/08/2008
350	Seniority	SEN SENVT	? ?	Union trademark file number: 569731 Date of publication of seniority: 08/08/2008
220	Date of application	AT	?	January 12, 1989
442	Date of publication	BT	?	April 15, 1989
151	Date of entry in the register	ET	?	Aug. 21, 1989
156	Extension of the term of protection	VBD	?	02/01/2019
730	Owner	INH	?	The Coca-Cola Company (n.d.Ges.d. State of Delaware), Atlanta Ga., US

Fig. 2.2 Coca-Cola script (DPMA)

A well-known word/figurative mark is the Coca-Cola script[9] of the Coca-Cola Company (see Fig. 2.2).

2.5.3 Figurative Mark

A figurative mark consists exclusively of graphic elements. Examples of a figurative mark are the Mercedes star[10] (see Fig. 2.3), the Adidas stripes of Adidas AG, or the bitten apple of Apple Corporation. A figurative mark should preferably be registered in black and white. This way, the figurative mark covers all common color designs.

2.5.4 Color Mark

In addition to the word mark, figurative mark, and word/figurative mark, there are other types of trademarks that, however, play little role in practice. One such exotic trademark is the color mark. With a color mark, an entire color is monopolized. A well-known color mark is the magenta[11] of Deutsche Telekom AG.

Deutsche Telekom AG can use its color mark to prohibit any provider of telecommunications services from using the color magenta. It was not easy for Deutsche Telekom AG

[9] DPMA, "https://register.dpma.de/DPMAregister/marke/register/1144921/DE", accessed June 20, 2021.

[10] DPMA, "https://register.dpma.de/DPMAregister/marke/register/2008220/DE", accessed June 20, 2021.

[11] DPMA, "https://register.dpma.de/DPMAregister/marke/registerHABM?AKZ=000212753&CURSOR=0", accessed June 20, 2021.

	Data	DB	?	DE
111	Registration number	RN	?	2008220
210	Old file number	AKZ	?	M70792
540	Brand representation	MD	?	
550	Brand form	MF	?	Figurative mark
551	Trademark category	MK	?	Individual trademark
220	Application date	AT	?	September 25, 1991
442	Date of publication	BT	?	February 15, 1992
151	Date of entry in the register	ET	?	Jan. 14, 1992
156	Extension of the term of protection	VBD	?	October 1, 2011
730	Owner	INH	?	Daimler AG, 70372 Stuttgart, Germany
750	Mailing address	ZAN	?	Daimler Brand & IP Management GmbH & Co. KG 063 - H512,70546 Stuttgart
	Version of the Nice Classification			NCL9
511	Nice class(es)	KL	?	**39**, 35, 36, 37, 38, 42, 43, 44
531	Picture class(es) (Vienna)	WBK	?	01.01.02, 24.11.25, 26.01.12
	File status	AST	?	Brand registered

Fig. 2.3 Mercedes star (DPMA)

to obtain the color mark. By now, it is almost impossible to have a color mark registered with the patent office. The prerequisite is that in a survey, more than half of the relevant public clearly associate the color with the applicant as the manufacturing or supplying company. It is easier to obtain a color mark in the B2B sector than in the B2C sector, since in this case the relevant public consists only of the customers or consumers of the specific market and the competitors. An example of a successful registration of a color mark in the B2B sector is the color red[12] of Hilti AG (see Fig. 2.4).

2.5.5 Sound Mark

There are also sound marks, in which sequences of sounds are monopolized for certain goods and services. For example, jingles can be protected. An example of a jingle protected as a trademark is the Telekom jingle[13] (see Fig. 2.5).

An important requirement for the registration of a trademark is graphical representability.[14] The purpose of this regulation is to ensure that the exact scope of protection can be determined in order to guarantee legal certainty. Only through graphical representation is

[12] DPMA, "https://register.dpma.de/DPMAregister/marke/registerHABM?AKZ=003425311&CURSOR=0", accessed June 20, 2021.

[13] DPMA, "https://register.dpma.de/DPMAregister/marke/register/399405917/DE", accessed June 20, 2021.

[14] Section 8(1) Trademark Act.

	Data set	DB	?	EM
111/210	Brand number	RN/ AKZ	? ?	003425311
540	Brand presentation	MD	?	
	Acquired distinctiveness			Yes
270	First language			German
270	Second language			English
550	Brand form	MF	?	Color mark
550	Trademark type EU trademarks	EUIPOMF	?	Color mark
591	Description of colors	FA	?	RED RAL 3020
551	Brand category	MK	?	Individual brand
220	Application date	AT	?	October 29, 2003
151	Date of entry in the register	ET	?	12/17/2008
730	Owner	INH	?	Hilti Aktiengesellschaft, 9494, Schaan, LI

Fig. 2.4 Color mark red of Hilti AG (DPMA)

	Database	DB	?	DE
111	Registration number	RN	?	39940591
210	File number	AKZ	?	399405917
540	Brand presentation	MD	?	
550	Brand form	MF	?	Sound mark
551	Brand category	MK	?	Individual trademark
220	Application date	AT	?	July 12, 1999
151	Date of entry in the register	ET	?	08/25/1999
156	Extension of the term of protection	VBD	?	01.08.2019
730	Owner	INH	?	Deutsche Telekom AG, 53113 Bonn, Germany
740	Representative	VTR	?	Hogan Lovells International LLP, 20095 Hamburg, DE
750	Mailing address	ZAN	?	Hogan Lovells International LLP, Alstertor 21, 20095 Hamburg
	Version of the Nice Classification			
511	Nice class(es)	KL	?	**38**, 9, 16, 25, 28, 35, 36, 37, 39, 41, 42
531	Picture class(es) (Vienna)	WBK	?	
	File status	AST	?	Trademark registered
180	Protection end date	VED	?	July 31, 2029
111 151	International registration			International registration number: IR729484 Registration date: 12/23/1999

Fig. 2.5 Jingle of Deutsche Telekom (DPMA)

	Database	DB	?	DE
111	Registration number	RN	?	39652556
210	File number	AKZ	?	396525563
540	Brand presentation	MD	?	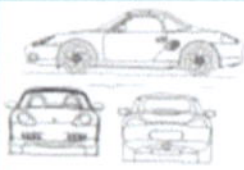
550	Brand form	MF	?	Three-dimensional trademark
551	Brand category	MK	?	Individual brand
220	Application date	AT	?	December 4, 1996
151	Date of entry in the register	ET	?	Jan. 20, 1997
156	Extension of the term of protection	VBD	?	Jan. 1, 2017
730	Owner	INH	?	Dr. Ing. h.c. F. Porsche Aktiengesellschaft, 70435 Stuttgart, Germany
740	Representative	VTR	?	UNIT4 IP Rechtsanwälte, 70174 Stuttgart, DE
750	Mailing address	ZAN	?	UNIT4 IP Rechtsanwälte, Jägerstr. 40, 70174 Stuttgart
	Version of the Nice Classification			NCL10
511	Nice class(es)	KL	?	9, 14, 16, 25, 28, 37
531	Picture class(es) (Vienna)	WBK	?	Jan. 18
	File status	AST	?	Trademark registered
180	Protection end date	VED	?	December 31, 2026
111	International registration			International registration number: IR676699

Fig. 2.6 3D mark Porsche Boxter (DPMA)

the correct information of the public about a registered trademark made possible. Sound marks can in particular be submitted in standard musical notation.[15] Sound marks are also referred to as "acoustic marks."

2.5.6 3D Mark

Another rare type of trademark is the 3D mark (three-dimensional mark). An example of a 3D mark is the Porsche Boxter[16] of Dr. Ing. h.c. F. Porsche AG (see Fig. 2.6).

A prerequisite for a 3D mark is that the shape is not typical for the goods for which the mark is to be registered. In addition, the three-dimensional shape must not be purely technically determined.[17] For example, Bahlsen GmbH & Co. KG and Mars Incorporated have registered 3D marks for special confectionery products.

[15] Section 11(2) Trademark Ordinance.

[16] DPMA, "https://register.dpma.de/DPMAregister/marke/register/396525563/DE", accessed June 20, 2021.

[17] *Federal Court of Justice, 15.12.2005, I ZB 33/04, NJW-RR, 2006, 1274* – Porsche *Boxster*.

2.5.7 Position Mark

Position marks are trademarks characterized by a specific position of a color on a product. A well-known example of a position mark is the red stripe on the heel of a shoe by Lloyd Shoes GmbH or the red sole of the shoe designs by designer Christian Louboutin.[18]

2.5.8 Thread Mark

Thread marks are particularly relevant for textile manufacturers. Textile manufacturers can weave a specific thread into their textiles to indicate origin. Target groups can thus identify the manufacturer of the clothing. With a thread mark, the manufacturer can signal directly on the textile that it is a product from their company.

Another example of the use of a thread mark is the use of a specially colored thread in a cable. The thread is seen by the electrician when stripping the cable, and the electrician can deduce the origin of the cable.

Thread marks play a minor role.

2.5.9 Collective Mark

With a collective mark, an association or organization can acquire a trademark.[19] The collective mark can be used by the members of the association or organization.

2.5.10 Other Types of Trademarks

Other types of trademarks include scent marks, motion marks, multimedia marks, and hologram marks.[20] These types of trademarks play no role in practice. The types of trademarks relevant in practice are the word mark, the figurative mark, and the word/figurative mark (see Fig. 2.7).

[18] In a decision by the ECJ, it was ruled that a red sole is eligible for trademark protection; ECJ, 12.6.2018, C-163/16, Industrial Property Protection and Copyright, 2018, 842—Christian Louboutin.

[19] Sections 97 to 106 Trademark Act.

[20] Section 12 Trademark Ordinance.

Fig. 2.7 Types of trademarks

Types of trademarks	
Important types of trademarks	**Exotic brand forms**
- Word mark	- Color trademark
- Figurative mark	- Sound mark
- Word/figurative mark	- 3D trademark
	- Position mark
	- Thread mark
	- Collective mark
	- Smell mark
	- Motion mark
	- Multimedia mark
	- Hologram mark

3 Word Mark versus Word/Picture Mark

The three most important types of trademarks are word marks, word/design marks, and design marks. A word mark consists of one or more words. An example of this is the trademark "After Eight"[1] (see Fig. 3.1).

A design mark consists only of a graphic, such as a logo. The bitten apple of Apple Corporation or the three stripes of Nike Corporation are design marks. A word/design mark is, in particular, the logo of the Coca-Cola Company. Usually, one faces the choice of registering either a word mark or a word/design mark. A design mark is advisable if a particularly memorable logo has been developed.

Example

Best Software GmbH is considering a trademark for its new financial accounting software product for small businesses. The CEO of Best Software GmbH suggests the trademark "FibuSoft." Everyone likes the trademark very much. The head of marketing suggests that a cloud as a symbol of the cloud would fit well. Preferably a cloud in the signal color red, since the trademark is supposed to be something very special. All members of the management team at Best Software GmbH find the "FibuSoft" trademark with a red cloud appealing, and it is decided that this is exactly the right trademark for the new product of Best Software GmbH.◄

[1] For example, the British word mark "After Eight" was registered on October 27, 1969, for Nice Class 30 in the United Kingdom. The registration number is UK 950237. The trademark owner is Nestlé (Société des Produits Nestlé S.A.) in Switzerland;

EUIPO, "https://www.tmdn.org/tmview/#/tmview/detail/GB500000000950237", accessed on June 20, 2021.

T. H. Meitinger, *Trademark Without a Lawyer*,
https://doi.org/10.1007/978-3-662-73127-7_3

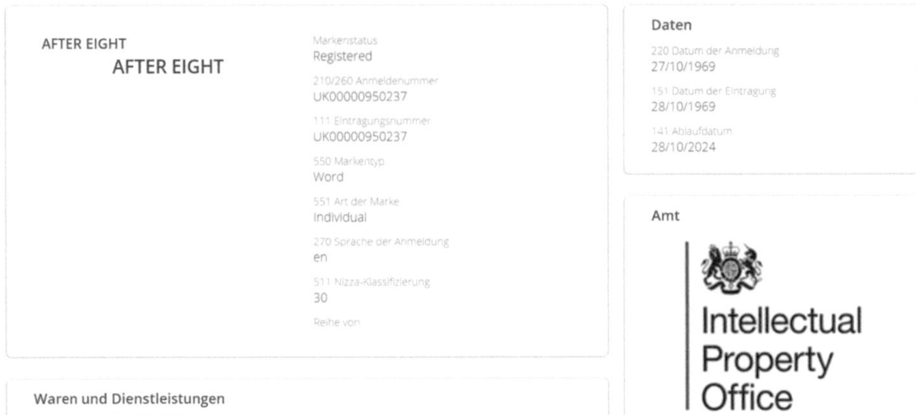

Fig. 3.1 Word mark "After Eight" (EUIPO)

In this example from Best Software GmbH, everything went wrong. The sign FibuSoft is descriptive of the product, namely the financial accounting software. A word mark FibuSoft would therefore have only a very limited scope of protection. The sign "FibuSoft" would, however, most likely not be registered for Class 9 (software and computers) by the patent office. By adding the red cloud, a word mark becomes a word/design mark, which alone reduces the scope of protection of the trademark. Adding a cloud as a graphic representation of the cloud will also not strengthen the trademark, since the concept "cloud" is descriptive for software. Overall, this results in a sign with a very limited inherent scope of protection, and the application for entry in the trademark register will likely be rejected by the patent office.

Example

Best Software GmbH has commissioned the development of a logo. The commissioned designer has created a stylish logo. The CEO of Best Software GmbH wants to have the logo protected as a trademark. He also wants the name of his own product to be included in the trademark. He argues that this way, you get protection for two things for the same price, namely the logo and the product name. The CEO of Best Software GmbH is therefore convinced that he should have a word/design mark protected.◄

The strength of a trademark is determined by its scope of protection. A word/design mark contains a text component and at least one graphic element. The multitude of elements in a word/design mark reduces the likelihood of confusion and thus diminishes the scope of protection of the trademark from the outset. A word/design mark therefore naturally has a small scope of protection. In comparison, a word mark generally has a broad scope of protection.

3.1 Word Mark

The essential element of a trademark is its scope of protection. The greater the scope of protection of a trademark, the more valuable the trademark is. A word/design mark with extensive text and graphic components has only a small scope of protection, since infringement only occurs if both the graphic and text components are realized by the infringing mark.

Example

Best Software GmbH has registered a word/design mark with an extensive graphic component and the word component "Pasquale" for software as a trademark. Now competitors appear with marks such as "Asquale," "Quale," or even "Pasquale." Unfortunately, due to the pronounced graphic elements of the word/design mark, no likelihood of confusion can be established, and Best Software GmbH is left powerless. With a word mark "Pasquale," the competitors could have been kept in check.◄

A word mark should be indicated exclusively in uppercase letters on the application form for trademark registration. This ensures, even for foreign applications, that the scope of protection of a registered word mark covers every form of writing. By capitalizing all the letters of the word mark, so-called internal capitalization is also included in the scope of protection of a word mark. In Germany, this is already ensured by indicating on the application form that it is a word mark. It is then irrelevant whether the word mark is written in lowercase, or whether the first letter is capitalized and the rest are lowercase.

Example

The word mark "PASQALE" is registered. In this case, all fonts and font sizes are included in the scope of protection, for example "pasqale", *"Pasqale",* "Pasqale", "Pasqale" or "PasQale". Color design is also generally included (see Fig. 3.2).

◄

Fig. 3.2 Scope of protection of a word mark

Scope of protection of a word mark		
Word mark	**Register as**	**Scope of protection**
"Pasqale"	"PASQALE"	Pasqale PASqale pasQale PASqale Pasqale Pasqale *Pasqale*

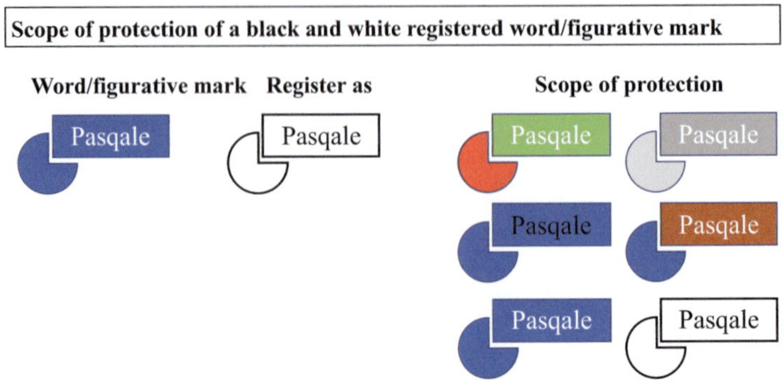

Fig. 3.3 Scope of protection of a word/design mark

A word mark should not be registered exclusively in uppercase letters if the unusual form of writing is precisely what makes it special. Such a feature could be an unusual font or alternating uppercase and lowercase letters. In this case, however, one is more likely to obtain a word/design mark than a word mark.

3.2 Word/Design Mark

A broad scope of protection for a word/design mark can be achieved if the word/design mark is registered in black and white or without colors. In this case, the scope of protection includes all word/design marks with different color schemes, as long as the color scheme is not unusual.

Fig. 3.3 shows a word/design mark with the text component "Pasqale." The trademark owner wants to use the word/design mark as shown on the left, with white lettering, a light blue rectangle, and a blue three-quarter circle. It is advisable to register the mark as shown in the center, in black and white. In this case, for example, all variants as shown under "Scope of protection" are included.

3.3 Word Mark or Word/Design Mark?

If you have a word/design mark, you should consider whether only the text component should be registered as a word mark instead of the word/design mark. The reason is that with a word mark, any use of the word mark for the registered goods and services by a third party, especially with any graphic elements, can be prohibited.

Therefore, unless there are compelling reasons for a word/design mark, for example because the word and graphic components harmonize in a special way, one should consider registering a word mark instead of a word/design mark. On the other hand, if the

Trademark form	Word component	Image component	Protected area
Word mark	Yes	No	large
Word/figurative mark	Yes	Yes	small

Fig. 3.4 Comparison of word mark and word/design mark

graphic element is so strong that it cannot be dispensed with, a design mark may be preferable to a word/design mark.

A word/design mark is sometimes recommended if the text component is not eligible for registration as a word mark. Lack of registrability may exist if the text component is descriptive for the required goods and services. In this case, a word/design mark with the descriptive text component and a distinctive graphic element can make the sign eligible for registration. However, this never results in a strong trademark.

Example

The words "günstige Autos" ("affordable cars") as a sign are not eligible for registration as a word mark for the goods cars, car seats, and car tires in Nice Class 12. However, a word/design mark with the words "günstige Autos" and a clever graphic symbol may be eligible for registration. In this case, however, the mark is not eligible for registration because of the text "günstige Autos," but because of the graphic element. The text component could then be omitted, as it does not establish any scope of protection, but at most reduces the scope of protection of the graphic element. The word/design mark only offers protection against marks that have the same graphic design. No protection is provided against competitors using the text component.◀

An advantage of a word mark over a word/design mark arises if the mark is to be graphically refreshed after a few years (facelifting).[2] There may be a desire for a "modern-looking" mark. With a word mark, this is usually not a problem, since all forms of writing are included. The mark can be refreshed in any way without failing to meet the statutory requirement for use of the mark.

The situation is different with a word/design mark. Facelifting can change the distinctive character of the mark. In this case, a new application for the word/design mark is required. The statutory requirement for use of the originally filed word/design mark is generally not met by a newly styled mark. For this reason, only a word mark can generally be used to create a strong, established mark (see Fig. 3.4).

[2] See Chap. 19 Facelifting and Relaunch.

3.4 Design Mark

A logo, in particular, can be protected as a design mark. If you have a distinctive, unusual logo, it may make sense to protect the logo with a design mark. If the logo is less distinctive and not imaginative, the corresponding design mark will generally be granted only a limited scope of protection.

Example

Best Software GmbH registers a cloud as a cloud symbol in red as a trademark for Class 9 for the goods software and computers at the German Patent Office. A cloud symbol is descriptive for software. Even if the mark is registered, it will generally be granted only a small scope of protection. If Bad Software GmbH registers and obtains a cloud symbol in green with a white background as a design mark for software, Best Software GmbH will probably not be able to successfully challenge Bad Software GmbH's mark. The scope of protection of Best Software GmbH's mark probably does not cover a different color of the cloud symbol.◀

4 Phases of a Trademark

The successive stages in the "life" of a trademark are presented in this chapter, and these stages are explored in greater depth in the following chapters. The "life" of a trademark can be extended indefinitely. The protection period for a trademark is 10 years each.[1] The protection period of a trademark can be extended any number of times by a further ten years each.[2] A trademark can therefore be maintained for an unlimited period by paying a renewal fee.

4.1 Naming

The first stage is naming. Successfully mastering this stage determines whether a valuable trademark can be acquired at all, because only an imaginative trademark can become a valuable trademark. Moreover, imaginative trademarks often have no or only minor issues with older trademarks, since trademark owners tend to protect descriptive rather than imaginative marks. The next major hurdle for a trademark, overcoming the absolute grounds for refusal—particularly lack of distinctiveness and infringement of the need to keep certain terms free for general use—generally poses no difficulty for imaginative trademarks. Finding the right trademark therefore determines whether the subsequent hurdles on the way to registration can be overcome smoothly or only with great effort.

[1] Section 47 paragraph 1 of the Trademark Act.

[2] Section 47 paragraph 2 of the Trademark Act.

T. H. Meitinger, *Trademark Without a Lawyer*,
https://doi.org/10.1007/978-3-662-73127-7_4

4.2 Registrability

Before a patent office enters a trademark into a trademark register, the trademark is examined for absolute grounds for refusal. The two most important hurdles here are distinctiveness and the need to keep certain terms free. A sign lacks distinctiveness if the relevant public does not recognize it as a trademark, i.e., as the designation of the manufacturer or provider. The need to keep certain terms free is infringed if the sign is descriptive for the goods and services for which it has been applied.

4.3 Search for Prior Rights—Likelihood of Confusion

Patent offices do not conduct a search for older trademarks that could give rise to a likelihood of confusion before registering a trademark.[3] Therefore, it is absolutely essential to conduct your own search for prior rights. Otherwise, it may happen that after several years—especially once the break-even point has long been surpassed and profits are being made—the owner of an older right demands that you cease further use of the trademark. In addition, the owner of the older trademark can charge you for the legal costs incurred in enforcing their rights. Such a disastrous event, which can threaten the very existence of a business, can only be reliably avoided by conducting a comprehensive search for prior rights yourself.

4.4 Filing a Trademark

A trademark application can be submitted by mailing the application documents or by filing them online. At the German Patent and Trademark Office (DPMA), application documents are also accepted by fax as an alternative.

4.5 Opposition

Patent offices do not check before registration whether a trademark infringes older rights. Therefore, patent offices do not protect the owner of an older trademark from a younger trademark that is likely to be confused with their older mark. As a remedy, the opposition procedure was established, which enables the owner of an older right to remove a younger, confusingly similar trademark from the register quickly and at low cost.

[3] An exception is the US Patent and Trademark Office (USPTO). The USPTO searches for prior rights before registering a trademark and informs the applicant of these prior rights.

4.6 Cancellation

After the opposition period has expired, a trademark can only be removed from the register through a cancellation procedure. A German trademark can be challenged by an official invalidity procedure before the patent office. Alternatively, a cancellation action can be filed with a regional court.

4.7 Trademark Monitoring

The owner of a trademark should regularly, at least once a month, look out for younger trademarks that are likely to be confused with their own. Younger, confusingly similar trademarks can lead to market confusion. In the worst case, your own good reputation may suffer due to the actions of the owner of the younger trademark, especially if they market inferior goods under the trademark.

4.8 Enforcement of a Trademark

A trademark must be enforced against confusingly similar, younger trademarks. To this end, the trademark infringer can be issued a warning or a preliminary injunction can be obtained against them. In addition, the rights of the trademark owner can be enforced by filing a lawsuit in a civil court.

4.9 Overview of the Phases

Figure 4.1 shows the chronological sequence of the phases of a trademark and refers to the chapters of the book in which the respective phase is described in detail.

Fig. 4.1 Phases of a trademark

Phases of a brand	
Phase 1: Naming	Chapter 6
Phase 2: Registrability	Chapter 7
Phase 3: Search for prior rights	Chapter 9
Phase 4: Registering the trademark	Chapter 13
Phase 5: Opposition	Chapter 14
Phase 6: Cancellation	Chapter 15
Phase 7: Trademark monitoring	Chapter 16
Phase 8: Enforcement of the trademark	Chapter 17

5 Effects of a Trademark

A trademark is an economic monopoly right. With a registered trademark, the trademark owner is permitted to prohibit third parties from using their trademark in an identical or confusingly similar manner. It provides its owner with a claim for injunctive relief to take action against trademark piracy. In addition, the trademark owner has a claim for damages against the infringer. Figure 5.1 lists all claims a trademark owner may assert against a trademark infringer.

If two registered trademarks come into conflict that are identical or confusingly similar for the respective registered goods and services, the principle of priority applies. The trademark with the earlier priority date grants its owner the right to prohibit the use of the younger trademark.

5.1 Claim for Injunctive Relief

An unlawful use of a trademark gives rise to a claim for injunctive relief by the trademark owner.[1] The claim for injunctive relief is available exclusively to the trademark owner. If there are several trademark owners, each one may assert the claim for injunctive relief.[2]

A claim for injunctive relief arises from an imminent first infringement (Erstbegehungsgefahr) or if there is a risk of recurrence (Wiederholungsgefahr). An imminent first infringement requires serious and tangible indications of an impending unlawful use of

[1] Section 14(2) of the Trademark Act or Article 9(2) of the EU Trademark Regulation.

[2] Each trademark owner may assert the claim for injunctive relief in accordance with Section 744(2) of the German Civil Code (BGB).

T. H. Meitinger, *Trademark Without a Lawyer*,
https://doi.org/10.1007/978-3-662-73127-7_5

Fig. 5.1 Effects of a trademark

Effects of a trademark
• Right to injunctive relief • Right to information • Right to submission and inspection • Right to compensation • Right to destruction and recall

the trademark.[3] If an imminent first infringement is to be assumed directly, the trademark owner is entitled to a precautionary claim for injunctive relief.[4] The imminent first infringement must be concretely apparent so that a well-founded legal assessment is possible.[5] An imminent first infringement can especially be assumed if the trademark infringer declares or by an action conveys that they believe they have a right to use the trademark.[6] An imminent first infringement is particularly assumed if the alleged trademark infringer has registered the disputed trademark for themselves at the Patent Office.[7]

A prerequisite for a risk of recurrence is that, after an initial trademark infringement has occurred, there is a factual presumption of further trademark infringements.[8] The risk of recurrence is not eliminated by the cessation of the initial trademark infringement. Ceasing business operations, dismissing the managing director, or paying damages also do not remove the risk of recurrence.[9] Only the submission of a cease-and-desist declaration subject to penalty ends the risk of recurrence.[10] The contractual penalty must be set at a level that leaves no doubt as to the seriousness of the infringer's promise to cease and

---

[3] Federal Court of Justice (BGH), 22.4.2010, I ZR 17/05—Pralinenform II.

[4] BGH, 17.3.1994, I ZR 304/91, Gewerblicher Rechtsschutz und Urheberrecht, 1994, pp. 530–532—Beta.

[5] BGH, 25.2.1992, X ZR 41/90, Gewerblicher Rechtsschutz und Urheberrecht, 1992, pp. 612–614—Nicola; BGH, 13.3.2008, I ZR 151/05, Gewerblicher Rechtsschutz und Urheberrecht, 2008, pp. 912–913—Metrosex.

[6] BGH, 31.5.2001, I ZR 106/99, Gewerblicher Rechtsschutz und Urheberrecht, 2001, pp. 1174–1175—Berühmungsaufgabe; BGH, 26.1.2006, I ZR 121/03, Gewerblicher Rechtsschutz und Urheberrecht, 2006, pp. 429–431—Schlank-Kapseln; BGH, 10.4.2003, I ZR 291/00, Neue Juristische Wochenschrift, 2003, pp. 3197–3198—Buchclub-Kopplungsangebot; OLG Hamburg, 7.2.2005, 3 W 14/05, GRUR-RR, 2005, p. 223—WM 2006.

[7] BGH, 22.1.2014, I ZR 71/12, Gewerblicher Rechtsschutz und Urheberrecht, 2014, p. 382—REAL-Chips; BGH, 13.3.2008, I ZR 151/05, Gewerblicher Rechtsschutz und Urheberrecht, 2008, p. 912—Metrosex; BGH, 14.1.2010, I ZR 92/08, Gewerblicher Rechtsschutz und Urheberrecht, 2010, p. 838—DDR-Logo.

[8] BGH, 31.7.2008, I ZR 21/06, Gewerblicher Rechtsschutz und Urheberrecht, 2008, pp. 1108–1110—Haus & Grund III; BGH, 28.3.1996, I ZR 39/94, Gewerblicher Rechtsschutz und Urheberrecht, 1996, pp. 781–783—Verbrauchsmaterialien.

[9] BGH, 30.4.2008, I ZR 73/05, Gewerblicher Rechtsschutz und Urheberrecht, 2008, pp. 702–706—Internet-Versteigerung III.

[10] OLG Frankfurt, NJOZ, 2007, p. 4312—Kollektivmarke Volksbank; BGH, 15.2.2007, I ZR 114/04, Gewerblicher Rechtsschutz und Urheberrecht, 2007, p. 871—Wagenfeld-Leuchte.

desist.[11] A claim for injunctive relief can be effectively enforced in particular by means of a warning letter, a preliminary injunction, or in court proceedings.

5.2 Right to Information

The trademark owner can only quantify their claim for damages if they receive information about the scope of the trademark infringement. Accordingly, they are entitled to a right to information against the infringer.[12] The respondent must specify exactly how many products were sold, at what price, and to whom. In addition, the respondent must disclose the source of the goods marked with the trademark, if they did not themselves affix the trademark.

5.3 Right to Submission and Inspection

The trademark owner is entitled to a right to submission and inspection against a third party if there is sufficient probability of a trademark infringement by the third party.[13] The trademark owner should have the opportunity to verify whether a trademark infringement has actually occurred. The right to submission and inspection can therefore be regarded as a preliminary claim that serves to establish the existence of further claims of the trademark owner.

The right to submission and inspection plays only a minor role in trademark law, because, unlike a process patent, which can be infringed behind factory walls, a trademark infringement is visible in the marketplace.

[11] BGH, 7.10.1982, I ZR 120/80, Gewerblicher Rechtsschutz und Urheberrecht, 1983, pp. 127–128—Vertragsstrafeversprechen.

[12] Section 19 of the Trademark Act; BGH, 29.9.1994, I ZR 114/84, Gewerblicher Rechtsschutz und Urheberrecht, 1995, p. 50—Indorektal/Indohexal; BGH, 26.11.1987, I ZR 123/85, Gewerblicher Rechtsschutz und Urheberrecht, 1988, pp. 307–308—Gaby.

[13] Section 19a of the Trademark Act.

Fig. 5.2 Limits of the effect of a trademark

Limits of the effect of a trademark
•Statute of limitations • Forfeiture •Exhaustion

5.4 Claim for Damages

Based on the information provided by the infringer, damages[14] can be calculated. Damages can be calculated in three ways.[15] The trademark owner may demand compensation for their loss[16] and that the lost profits be transferred to them[17]. Alternatively, the trademark owner may demand the surrender of the infringer's profits (account of profits).[18] In the past, the account of profits was rarely chosen, as infringers would regularly understate their profits. The Federal Court of Justice has recently put a stop to this abusive practice. The third option is to calculate damages based on a notional license fee (license analogy).[19] The choice of calculation method is up to the trademark owner.

5.5 Claim for Destruction and Recall

The trademark owner has a claim for destruction regarding goods in the possession or ownership of the infringer that are unlawfully marked with the trademark owner's mark.[20] The claim for destruction also extends to materials and equipment owned by the infringer that are primarily used for unlawful marking.[21] The trademark owner has a claim for recall of goods unlawfully marked with the trademark by the infringer.[22] A claim for destruction and recall exists only if the enforcement of the claims is proportionate.

5.6 Limits of the Effect of a Trademark

Figure 5.2 lists the limits of the effect of a trademark.

---

[14] Section 14(6) sentence 1 of the Trademark Act.

[15] BGH, 2.2.1995, I ZR 16/93, Gewerblicher Rechtsschutz und Urheberrecht, 1995, p. 349—Objective calculation of damages.

[16] Sections 249 ff. BGB.

[17] Section 252 BGB.

[18] Section 14(6) sentence 2 of the Trademark Act.

[19] Section 14(6) sentence 3 of the Trademark Act.

[20] Section 18(1) sentence 1 of the Trademark Act.

[21] Section 18(1) sentence 2 of the Trademark Act.

[22] Section 18(2) of the Trademark Act.

5.6.1 Limitation Period

Claims arising from a trademark, including the claim for injunctive relief, the right to information, the right to submission and inspection, the claim for damages, and the claim for destruction and recall, become time-barred under the standard limitation period of the German Civil Code after three years.[23]

5.6.2 Forfeiture

Forfeiture means that the trademark owner, through their own fault, can no longer enforce their rights arising from the trademark. The concept of forfeiture in trademark law was developed by the courts based on the principle of prohibition of abusive exercise of rights.[24] Forfeiture results from a balancing of interests between the trademark owner and a bona fide infringer.

The prerequisites for forfeiture are that the infringer has used the trademark undisturbed and in good faith over a longer period of time. In addition, the infringer must have created significant value through the use of the trademark, which, according to the principle of good faith[25] must be preserved for them. Furthermore, this value must have arisen as a result of the conduct or inaction of the trademark owner.[26]

What is particularly detrimental for the trademark owner about forfeiture is that it can have a spillover effect on other, younger trademarks. The owner of a younger trademark

---

[23] Section 20 sentence 1 of the Trademark Act in conjunction with Section 195 BGB.

[24] Section 242 BGB.

[25] Section 242 BGB.

[26] BGH, 31.7.2008, I ZR 171/05, Gewerblicher Rechtsschutz und Urheberrecht, 2008, pp. 1104-1107—Haus & Grund II; BGH, 31.7.2008, I ZR 21/06, Gewerblicher Rechtsschutz und Urheberrecht, 2008, p. 1108—Haus & Grund III; BGH, 21.7.2005, I ZR 312/02, Gewerblicher Rechtsschutz und Urheberrecht, 2006, p. 56—BOSS-Club; BGH, 6.5.2004, I ZR 223/01, Gewerblicher Rechtsschutz und Urheberrecht, 2004, pp. 783-785—Neuro-Vibolex/Neuro-Fibraflex; BGH, 15.2.2001, I ZR 232/98, Gewerblicher Rechtsschutz und Urheberrecht, 2001, pp. 1161-1163—CompuNet/ComNet; BGH, 24.6.1993, I ZR 187/91, Gewerblicher Rechtsschutz und Urheberrecht, 1993, pp. 913-914—KOWOG; BGH, 30.11.1989, I ZR 191/87, Gewerblicher Rechtsschutz und Urheberrecht, 1992, p. 329—Ajs-Schriftenreihe; BGH, 7.6.1990, I ZR 298/88, Gewerblicher Rechtsschutz und Urheberrecht, 1990, p. 1042—Datacolor; BGH, 2.2.1989, I ZR 183/86, Gewerblicher Rechtsschutz und Urheberrecht, 1989, p. 449—Maritim; BGH, 26.5.1988, I ZR 227/86, Gewerblicher Rechtsschutz und Urheberrecht, 1988, p. 776—PPC; BGH, 27.6.1980, I ZR 70/78, Gewerblicher Rechtsschutz und Urheberrecht, 1981, p. 66—MAN/G-Man; BGH, 23.9.1992, I ZR 251/90, Gewerblicher Rechtsschutz und Urheberrecht, 1993, p. 151—Universitätsemblem; BGH, 22.11.1984, I ZR 101/82, Gewerblicher Rechtsschutz und Urheberrecht, 1985, p. 389—Familienname; BGH, 10.11.1965, Ib ZR 101/63, Gewerblicher Rechtsschutz und Urheberrecht, 1966, p. 623—Kupferberg; BGH, 30.1.1963, Ib ZR 118/61, Gewerblicher Rechtsschutz und Urheberrecht, 1963, pp. 478-481—Bleiarbeiter.

can rightly argue that the owner of the older trademark has already tolerated the use of similar, younger trademarks and that one must therefore be able to rely on this also applying to another younger trademark. It is therefore important for the trademark owner to keep the trademark register "clean." Trademark monitoring is necessary to avoid having to tolerate the erosion of one's own trademark rights.

5.6.3 Exhaustion

If a product is marked with the trademark owner's mark with their consent and placed on the market in Germany or in another member state of the European Union or in a contracting state of the Agreement on the European Economic Area, exhaustion occurs. In this case, the trademark owner's right to control the use of the trademark for the marked goods is exhausted.[27]

Example

Best Software GmbH has a registered trademark "Pasqale" in Germany. In France, Best Software GmbH has no trademark protection. In France, Best Software GmbH sells a product marked with the "Pasqale" trademark. Best Software GmbH cannot prohibit the export and sale of this product, sold in France, in Germany. The trademark protection of Best Software GmbH has no effect in Germany with respect to this product.◀

Exhaustion prevents the trademark owner from exerting influence over the further distribution of a product marked with their trademark. This puts a stop to the trademark owner's complete control over the distribution channel. The trademark owner's rights are therefore exhausted by the one-time consent to the marking with the trademark.

[27] Section 24(1) of the Trademark Act.

6 Finding a Brand Name

Before a trademark is submitted for registration at a patent office, several important points need to be clarified. First, a good brand name should be found. The goods and services for which the trademark is to be registered must be determined. After that, it should be checked whether the trademark is eligible for registration. Finally, research must be conducted to determine whether there are any earlier trademarks that could lead to a likelihood of confusion.

Finding the right name is the most important step on the path to a valuable brand. Only with a sign that has a high recognition value can a valuable brand be built. A sign like "günstigAutokaufen" ("buyCarCheap") for cars will have a hard time achieving high recognition value. "günstigAutokaufen" can easily be confused with "günstigesAuto" ("cheapCar"), "Auto günstig kaufen" ("buy car cheaply"), or "mein günstiges Fahrzeug" ("my cheap vehicle"). In contrast, a brand like "Zeus" or "Alhambra" for cars will be able to build a good recognition value more easily.

The fundamental steps in finding a name for a brand are a market analysis, brainstorming, eliminating problematic signs, and a selection process. Figure 6.1 shows the various steps in the naming process.

6.1 Market Analysis

A market analysis serves to analyze the brands in the industry for which your own brand is being developed. This provides an initial orientation as to what types of signs are common and have proven successful. You will also find brands that seem rather awkward.

T. H. Meitinger, *Trademark Without a Lawyer*,
https://doi.org/10.1007/978-3-662-73127-7_6

Fig. 6.1 General method for finding a name

General method
• Market analysis • Brainstorming • Delete problematic characters • Selection process

Of course, no foreign brand should be copied. Rather, the goal is to develop a feel for the particularities of the market. It is also important that your own brand is clearly differentiated in character from the brands of competitors. Only in this way can the brand embody the position of your own company and stand out against competitors. Your own brand can then be "charged" with your own values.

6.2 Brainstorming

Brainstorming is a proven creativity technique for finding a name for a product or company. Brainstorming is carried out in a group of four to eight participants. It is important that the brainstorming takes place in a pleasant, relaxed atmosphere. It can be disadvantageous if the "boss" is present. It is advisable that the participants are from the same hierarchical level and are therefore equals.

A basic principle of brainstorming is that quantity comes before quality ("quantity over quality"). As many ideas as possible should be collected. Criticism is forbidden. The flow of ideas should not be interrupted. No selection is made at this stage. Ideas should also not be discussed. However, ideas can be picked up and further developed. Unusual ideas are not only allowed but expressly encouraged.

6.3 Eliminating Problematic Signs

Signs that do not sound phonetically harmonious or are difficult to pronounce should not be pursued further. Signs with difficult spelling should also be eliminated. The same applies to signs that jump on a current fashion trend. There is a high risk that after the trend ends, the brand will seem outdated.

The brand should not have a negative meaning in the languages of the future target countries. It should also sound pleasant in the respective foreign language, and the brand should be able to be written in that language. At the very least, these requirements should be met for the world language English.

Brands that are too similar to those of competitors[1] should be discarded. Brands that seem artificial or uninspired are also not suitable candidates for trademark registration.

[1] Competing companies or competitors are alternatively referred to as "market companions".

Mitsubishi had problems with negative connotations with "Pajero". In Spanish-speaking countries, the word "Pajero" is an insult. Mercedes-Benz had to withdraw its brand "Bensi" for China because "Bensi" sounds ominous in Chinese.

6.4 Selection Process

The developed brands can be subjected to a field test, for example by asking friends, acquaintances, and relatives for their opinions. Brands that do not receive a positive spontaneous response or require explanation can thus be eliminated.

6.5 Recommended Method

An alternative method, which has often proven itself in practice, consists of four steps: focusing the message, forming associations, combining the associations, and fine-tuning. Figure 6.2 provides an overview of the individual steps in the naming process.

6.5.1 Focusing the Message

Based on an analysis of the market and competition, two messages should be identified that the brand is intended to embody. These messages should lead to a unique selling proposition for your own business activity. In particular, you can think about the desired characteristics of the product or service.

It is important to focus in this step. Deliberately limit yourself to just two messages to avoid getting sidetracked. These two messages lead to the targeted core competency of the goods and services identified by the brand. The finished brand should reflect the core competency.

A company that claims to have ten core competencies probably has none at all. A core competency must be purposefully built and constantly developed. This requires extensive resources. Trying to build ten core competencies at once will usually exceed the available resources. It is advisable to limit yourself to two core competencies.

Fig. 6.2 Recommended method for finding a name

Recommended method
• Focusing the message • Forming associations • Connect associations • Fine-tuning

6.5.2 Forming Associations

Synonyms and similar terms for the two messages are sought. The rule "quantity over quality" applies. As many terms as possible should be found. Unusual terms and imaginative associations are especially welcome. For each of the two messages, you will end up with a list of synonyms, similar terms, and associations.

6.5.3 Combining Associations

The terms from the two lists are combined. Any combinations can be made. Variations that include additional elements not on the lists are expressly encouraged. There are no limits to creativity. It is not necessary to slavishly stick to the existing synonyms or associations. Here, too, "quantity over quality" applies. As many variations as possible should be generated. Criticism or selection does not take place at this stage.

6.5.4 Selection and Fine-Tuning

The final step is selection. The generated combinations should be critically examined. Variations that are very long, difficult to write, or pronounce are discarded. Variations that sound nonsensical or uninspired should also be eliminated.

Developed terms can still be refined. Parts of the brand may be omitted to arrive at a catchy sign. A final selection can be made with a small field test involving friends and acquaintances.

6.6 Summary

Finding the right name is the first and most important step on the path to a valuable brand. Two methods have been presented to accomplish this step. There is nothing to prevent both methods or even additional methods from being applied in parallel.

It is not necessary to commit to a single brand immediately after the naming process. It makes sense to develop several, for example three, brand names through the naming process and subject them to trademark law examinations. The trademark law examinations primarily include checking for absolute grounds for refusal and for likelihood of confusion with earlier rights. Especially when checking for earlier rights, developed brand names may have to be discarded.

A sensible approach is therefore to apply both of the presented naming methods. In the selection process, arrive at a small group of three to five brand names and subject these to trademark law examinations.

7 Obstacles to Registration

Before entering a trademark into the trademark register, the Patent Office examines exclusively whether the trademark overcomes the so-called absolute grounds for refusal.[1] Signs that violate public order or accepted principles of morality are not registered as trademarks. In addition, signs that contain state coats of arms, national flags, official emblems, coats of arms of a domestic locality or a domestic municipal or other local authority, official inspection or certification marks, are not entered into the Patent Office's trademark register. Figure 7.1 provides an overview of the grounds for refusal.

7.1 Lack of Distinctiveness

A sign eligible for registration must be distinctive. Distinctiveness exists if the relevant public recognizes that the sign is intended to indicate the origin of the goods or services. The relevant public comprises the trade and the attentive, reasonable average consumer. Words such as "Super," "great," or "fantastic" lack distinctiveness. These words are not perceived by the relevant public as an indication of origin, but merely as promotional for the goods or services. Such signs cannot be registered as trademarks in the Patent Office's register.[2]

---

[1] Section 8 Trademark Act.

[2] Section 8(2) No. 1 Trademark Act.

T. H. Meitinger, *Trademark Without a Lawyer*,
https://doi.org/10.1007/978-3-662-73127-7_7

Fig. 7.1 Absolute grounds for refusal

Grounds for refusal
• Lack of distinctiveness • Need to keep free • Risk of deception • Public order and morality • National coats of arms and flags • Official test and certification marks • Bad faith

7.2 Need to Keep Free

Signs that may serve to designate the kind, quality, quantity, intended purpose, value, geographical origin, time of production of the goods or provision of the services, or other characteristics of the goods or services, are excluded from registration.[3] The rationale is that every market participant must be able to describe their goods and services in terms of their characteristics. It should be noted that the need to keep free applies specifically to certain goods and services.

A sign that is descriptive is excluded from registration in the trademark register. An exception exists for a trademark that has acquired distinctiveness through use. Acquired distinctiveness means that at least 80% of the relevant public recognizes the sign as the mark of origin of the respective company.

7.3 Risk of Deception

Signs that are likely to deceive the public, in particular regarding the nature, quality, or geographical origin of the goods or services, are not registered as trademarks.[4] For example, a sign that suggests the company operates nationwide, although the company cannot in any way fulfill this claim, constitutes such a deception.

7.4 Public Order and Accepted Principles of Morality

A violation of a law does not automatically constitute a violation of public order. Only a breach of the fundamental principles of German law fulfills this criterion. A violation of public order should be determined with restraint, as it is an exceptional circumstance.

[3] Section 8(2) No. 2 Trademark Act.

[4] Section 8(2) No. 4 Trademark Act.

A violation of accepted principles of morality arises if there is a gross offense to taste in moral, political, or religious terms.[5] The offense to taste must be perceived as such by the majority of the intended audience for the sign.

7.5 State Coats of Arms, National Flags, and Official Inspection and Certification Marks

A trademark may not contain a state coat of arms, national flags, or official inspection and certification marks. Otherwise, registration in the trademark register is excluded.[6]

7.6 Bad Faith Application

A trademark that has been applied for with the sole intention of obstructing a third party will not be registered as a trademark.[7] The prerequisite is that the Patent Office is aware of the earlier trademark.

There are cases where a product is already on the market under a trademark, without the trademark having been registered with the Patent Office, and a third party applies to register this trademark for exactly these goods and services. In this case, a bad faith trademark application may exist, which can be challenged by a nullity action[8] before the Patent Office or by an action for cancellation before the Regional Court.

Example

Best Software GmbH sells its software under the trademark "Pasquale." Best Software GmbH has not registered a trademark. One year later, Bad Software GmbH registers the trademark "Pasquale" for class 9 for computers and software and class 42 for software development and also uses the trademark. Because of this use, there is no bad faith application. Bad Software GmbH is entitled to prohibit Best Software GmbH from continuing to use the trademark "Pasquale" for software.◄

The owner of a trademark can prohibit any third party from using their trademark for the registered goods and services. Bad Software GmbH has the right to prohibit Best Software GmbH from using the trademark "Pasquale" for software. In this case, Best Software GmbH is left with only two options: to cease using the trademark "Pasquale" or

[5] Federal Court of Justice, 18.9.1963, Ib ZB 21/62, Industrial Property Protection and Copyright Law, 1964, p. 136—Swiss Cheese.

[6] Section 8(2) Nos. 6 and 7 Trademark Act.

[7] Section 8(2) No. 14 Trademark Act.

[8] Section 53(1) sentences 1 and 2 in conjunction with Section 50(1) Trademark Act.

to challenge Bad Software GmbH's trademark registration as a bad faith application. For this, a cancellation request must be filed with the Patent Office, demonstrating that the trademark was registered solely to obstruct one's own business activities. A cancellation procedure is a lengthy process. One should expect a duration of approximately one year.

A prerequisite for a successful cancellation procedure due to a bad faith trademark application is that the trademark was registered purely with the intention to obstruct. Proving exclusivity is difficult here, since Bad Software GmbH is using the trademark "Pasquale." Therefore, it cannot be proven that Bad Software GmbH acted solely with the intention to obstruct. A cancellation procedure will likely be unsuccessful.

Alternatives to a cancellation procedure are a preliminary injunction or an action for unfair competition. However, the same principles apply. It must also be shown that Bad Software GmbH has no legitimate interest of its own, but registered the trademark solely with the intention to obstruct. Such proof is hardly possible in the case of a trademark that is being used.

A different situation would arise if it were not Best Software GmbH, but Pasquale Software GmbH that had distributed its software under the name "Pasquale."

Example

Pasquale Software GmbH distributes software under the trademark "Pasquale." The trademark "Pasquale" was not registered as a trademark. Bad Software GmbH applies to register the trademark "Pasquale" with the Patent Office. In this case, there is already a company name right of Pasquale Software GmbH prior to the application for the trademark "Pasquale" by Bad Software GmbH. Due to the earlier company name right, the trademark "Pasquale" of Bad Software GmbH can be challenged. For example, an opposition can be filed with the German Patent Office.◄

A prerequisite for opposition proceedings based on a company name right is that it can be proven that a relevant company name right existed before the filing date of the opposed trademark. The proof must show, in particular, that there was nationwide use with sufficient turnover in the goods and services for which the opposed trademark was registered.

7.7 Registration by Acquired Distinctiveness

The grounds for refusal of lack of distinctiveness and infringement of the need to keep free are disregarded if the trademark in question has acquired distinctiveness through use.[9]

Acquired distinctiveness must be distinguished from reputation. Reputation is sufficient for a trademark that is in principle eligible for registration to acquire trademark rights

[9] Section 8(3) Trademark Act.

Fig. 7.2 Reputation versus acquired distinctiveness

Market recognition versus market penetration	
Trademark is fundamentally registrable	Trademark is fundamentally not registrable
Market recognition leads to a trademark based on use	Market penetration overcomes lack of registrability
20-25% awareness in the relevant trade circles	Over 50% awareness in the relevant trade circles
Regional awareness is sufficient	Nationwide recognition required

through use.[10] Reputation does not require that the trademark is known to a majority of the relevant public. Acquired distinctiveness is required for a trademark to be registered even though it is not eligible for registration. For acquired distinctiveness, it is necessary that a majority of the relevant public knows the trademark. Acquired distinctiveness is therefore present if more than 50% of the relevant public associate the trademark with the manufacturer or provider of the designated goods.

In addition, for acquired distinctiveness, as a rule, recognition throughout the entire territory of Germany is required. For reputation, local recognition is usually sufficient (see Fig. 7.2).

[10] Section 4 No. 2 Trademark Act.

8 Risk of Confusion

Before registering a trademark, it should be checked whether there are any prior rights that could lead to a likelihood of confusion with the trademark. If a trademark is used to build a business activity without checking for the likelihood of confusion, a disaster can ensue. The business activity is started and, for example, after seven years, the owner of an earlier right appears and prohibits further use of the trademark. In addition, they may assert a claim for the surrender of the infringer's profits. A nightmare scenario and possibly an abrupt end to the business activity.

The German Patent and Trademark Office (DPMA) does not search for earlier trademarks whose scope of protection would be infringed by a trademark to be registered. Before entering the trademark in the register, the German Patent and Trademark Office only checks whether the absolute grounds for refusal are violated.

The EUIPO (European Union Intellectual Property Office) searches for prior rights and prepares a search report.[1] The owners of the prior rights identified in the search are informed after the publication of the trademark application.[2] They are thus given the opportunity to file an opposition against a later trademark.[3] The search report is also sent to the applicant of the later trademark upon request.[4] No additional costs are incurred by the applicant.

To identify similar (or even identical) earlier trademarks, a so-called similarity search should be conducted before applying for a trademark. In a similarity search, only those earlier trademarks that have been registered for similar or identical goods and services

[1] Article 43(1) of the European Union Trade Mark Regulation.

[2] Article 43(7) sentence 1 of the European Union Trade Mark Regulation.

[3] Article 46(1) of the European Union Trade Mark Regulation.

[4] Article 43(6) of the European Union Trade Mark Regulation.

T. H. Meitinger, *Trademark Without a Lawyer*,
https://doi.org/10.1007/978-3-662-73127-7_8

Fig. 8.1 Examination of likelihood of confusion

Examination of the likelihood of confusion
•Similarity of the goods and services •Similarity of the marks – Phonetic similarity – Visual similarity – Conceptual similarity •Scope of protection of the earlier trademark

should be considered. For example, if you want to register a trademark for food products, earlier trademarks for cars are likely to be irrelevant. Alternatively, a search can be conducted for trademarks of selected third parties for specified goods and services. This enables a competitive analysis.

When conducting a trademark search for a German trademark, searches must be made for German trademarks, for European Union trademarks, and for international registrations designating Germany or the European Union.

If you wish to register a European Union trademark, searches must be made for European Union trademarks, for national trademarks of all EU member states, and for international registrations designating the European Union or a member state of the European Union.

A likelihood of confusion exists if the relevant public cannot distinguish between two trademarks. If a consumer hears or sees a first trademark of a first trademark owner and then, for example an hour later, a second trademark of a second trademark owner, and believes that the first and second trademark owners are the same person or the same company, then there is a likelihood of confusion.

The prerequisite for a likelihood of confusion is not only that the two trademarks are identical; sufficient similarity is enough. It is also not necessary for the same goods and services to be offered under both trademarks. It is sufficient if the goods and services are so similar that the consumer assumes they come from the same provider. A likelihood of confusion means that a later trademark encroaches on the scope of protection of an earlier trademark.

Figure 8.1 shows the basic examination scheme. The first step is to check the similarity of the goods and services for which the two trademarks are registered or applied for. If similarity of the goods and services is excluded, there can be no likelihood of confusion and the examination is concluded.

The next step is a comparison of the trademark representations. If there is no similarity between the trademark representations, a likelihood of confusion is excluded. Otherwise, it must be examined whether the similarity of the goods and services and the similarity of the trademark representations are sufficient for a likelihood of confusion. In addition, it is checked whether the earlier trademark is assumed to have an extended scope of protection due to intensive use. In this case, a likelihood of confusion is more likely to be assumed. Otherwise, an average distinctiveness of the earlier trademark is assumed. For the later

Fig. 8.2 Criteria for similarity of goods and services

Similarity of the goods and services

- Nature of the goods and services (composition, function, quality, appearance, and value/price of the goods or services)
- Intended use (area of application)
- Method of use (use of the goods or services)
- Complementary goods and services
- Competing goods and services
- Distribution channels (distribution channels and points of sale)
- Regular origin of the goods and services

trademark, no extended scope of protection is assumed, since intensive use could not yet have taken place with the later trademark.

A special case is double identity. In this case, both the goods and services and the trademark representations are identical. A likelihood of confusion is to be affirmed without hesitation.

8.1 Similarity of Goods and Services

Various aspects are considered when determining the similarity of goods and services. As a general rule, similarity of goods cannot be inferred simply because the goods or services in question belong to the same Nice class. This can be illustrated by class 9, for example, which includes both software and sunglasses and contact lenses. The classification of goods and services does not directly result in similarity of goods or services. The similarity of goods and services is determined by their nature, intended purpose, use, whether they are competing or complementary goods and services, distribution channels, relevant public, and the usual origin of the goods and services (see Fig. 8.2).[5]

8.1.1 Nature of Goods or Services

The nature of a good or service can be derived from factors such as composition, mode of operation, quality, appearance, and value of the good or service. Among these factors, particular importance should be attached to those that are of special significance from the perspective of the relevant public (see Fig. 8.3).

For example, for alcoholic beverages, the **composition,** especially the alcohol content, is of particular importance. The appearance, such as the shape of the bottle in which the alcohol is filled, is of little significance. On the other hand, for clothing, appearance is

[5] ECJ, 29.9.1998, C-39/97, Gewerblicher Rechtsschutz und Urheberrecht, 1998, p. 922—Canon.

Fig. 8.3 Nature of goods and services

Type of goods and services

- Composition of the goods (raw materials and components)
- Functionality (mechanical, optical, electrical, biological, or chemical functionality)
- Nature (solid, liquid, gaseous)
- Appearance
- Value or price of the goods or services

important, for example as women's or men's clothing. The composition of the clothing usually plays only a subordinate role.

The **mode of operation** of a product can be divided into mechanical, optical, electrical, biological, or chemical operation. For example, there are bending machines that operate mechanically and can bend sheet metal. On the other hand, washing machines operate chemically, as the desired effect is achieved by chemical means, especially detergents. From the perspective of mode of operation, bending machines and washing machines are therefore dissimilar.

With regard to **quality,** particular attention should be paid to the usual quality, especially whether the product is typically solid, liquid, or gaseous. For example, beverages are liquid and baked goods are solid. Quality is a less important indicator of similarity or dissimilarity of goods.

For toys, **appearance** is decisive. The appearance must be such that it is attractive to the child. The material from which the toy is made is secondary, as long as the material is not harmful to the child's health.

A significantly **different price of a good or service** should be interpreted with caution as an indication of dissimilarity of goods or services. For example, buying a new car and car leasing are similar services, even though the customer faces significantly different prices for the product or service.

8.1.2 Intended Purpose

The intended purpose refers to the intended use of a good or service, i.e., the use for which the good or service is usually offered. The regular intended purpose is meant to satisfy the customer's need.

The intended purpose is an important aspect in assessing the similarity of goods and services. It determines the intended use of the goods and services. If the intended purpose is similar, similarity of the goods and services will generally be found, as in this case the goods or services are in competition with each other. An example is shampoo and shower gel, both of which serve personal care. Shampoo and shower gel can be considered similar goods.

However, one should not rely blindly on the intended purpose. Toothpaste and shower gel have the same intended purpose, namely cleaning and hygiene of the body. Nevertheless, toothpaste and shower gel can only be considered to have a low degree of similarity.

8.1.3 Method of Use

The method of use describes the procedure by which the good or service is applied or used to achieve its intended purpose. The method of use thus leads to the satisfaction of the customer's need for the good or service.

The method of use for a lawnmower is pushing it over a lawn, thereby achieving the intended purpose of cutting the grass. A stroller can also be pushed over a lawn, thereby achieving the intended purpose of transporting a sleeping baby. However, the same method of use—pushing—does not mean that a lawnmower and a stroller are similar goods.

8.1.4 Complementary Goods or Services

Goods or services can complement each other and thus constitute complementary products. If the goods or services can only be used together to achieve the intended purpose, there is a high degree of dependency between the goods and services, and similarity is to be assumed. An example is cigarette tobacco and rolling papers for hand-rolled cigarettes. Another example could be towels and beach articles.

8.1.5 Competing Goods or Services

Competing goods and services are aimed at the same customer groups and are economically interchangeable. Competing products are therefore in competition with each other. The relevant public assumes that competing goods or services may come from the same source. Examples of competing products are butter and margarine, gold jewelry and costume jewelry, or new car purchase and car leasing (see Fig. 8.4).

8.1.6 Distribution Channels

A manufacturer can distribute a product through retail or wholesale. There is also network marketing, where the product is sold directly to customers and resellers. The resellers may in particular have been recruited by the seller. If goods are distributed through the same distribution channels, the relevant public is more likely to perceive the goods as similar.

Competing goods and services		
Economically interchangeable, aimed at the same customers, same intended use		
Examples:		
Butter	versus	Margarine
Purchase of a new car	versus	Car leasing
Gold jewelry	versus	Fashion jewelry

Fig. 8.4 Competing goods and services

Points of sale include, for example, supermarkets, drugstores, and department stores. Supermarkets offer a wide variety of goods, from food and writing utensils to adhesives. For this reason, the same point of sale is not a strong argument for similarity of goods. However, if the products to be compared are mainly offered in specialty stores that only offer a limited range of goods, similarity of goods is more likely to be assumed.

8.1.7 Relevant Public

The relevant public includes, in particular, current and potential customers. If there is an overlap between the customer groups for different goods and services, similarity of goods and services can be assumed. However, it should not be overlooked that the same customers may have a need for different goods and services. The identity or overlap of customer groups therefore does not automatically lead to similarity of goods or services. Lawnmowers and strollers may be aimed at the same customer group, namely young fathers, and yet they are not similar goods.

However, different customer groups can be interpreted as an indication of dissimilar goods or services. If, in particular, the first goods or services are aimed at end consumers and the second goods or services at business customers, the goods or services are usually dissimilar.

8.1.8 Usual Origin of Goods or Services

The same usual origin is a strong indication of similarity of goods or services. The same usual origin exists if the relevant public is accustomed to receiving the different goods or services from the same or economically related companies.

However, the fact that large conglomerates offer both a first and a second good or service does not automatically mean that the first good or service is similar to the second. An example is the Procter & Gamble Company, which manufactures and sells, for example, detergents, coffee machines, and baby diapers. Despite the same origin, detergents, coffee machines, and baby diapers are dissimilar goods.

8.2 Similarity of a Good with a Service

When assessing the similarity of a good with a service, the same principles generally apply as when comparing a first good with a second good or a first service with a second service.

Goods and services are fundamentally different. Goods are products whose sale involves a transfer of ownership. In contrast, services are intangible and cannot be stored. Despite the differences between goods and services, a good can have a complementary or competing relationship to a service. The service of maintenance and servicing of a copier can be regarded as a complementary service to the good copier. The service of car leasing can compete with the good new car.

8.3 Similarity of Signs

An examination of similarity of signs must be comprehensive, meaning that similarity must be assessed in terms of visual appearance, sound, and conceptual meaning. To affirm similarity of signs, it is sufficient if similarity exists in visual, phonetic, or conceptual terms.

A clear difference in one respect can mean that, despite similarity in another respect, a likelihood of confusion is to be denied. An example is a word/device mark and a word mark, where a graphic element of the word/device mark is so dominant that phonetic similarity is outweighed by a high degree of visual dissimilarity. A word/device mark can therefore be dissimilar to a word mark with the same textual component.

On the other hand, even a very slight similarity in visual, phonetic, and conceptual terms, which individually would not be sufficient for similarity, can, when considered as a whole, lead to a likelihood of confusion.

When assessing the visual and conceptual similarity of two signs, it must be considered whether the relevant public is able to correctly perceive the visual and conceptual content of the signs; in assessing the likelihood of confusion, it is important to consider how the visual and conceptual elements of the signs are perceived. An assessment of similarity can only be made in the context of the understanding of the relevant public.

In some situations, the textual elements of a trademark should be emphasized over the graphic elements, as trademarks are used in spoken form in the vast majority of cases. In such cases, even strong graphic elements alone do not help to distinguish a word/device mark from a word mark with the same textual component.

8.3.1 Overall Impression

When assessing similarity of signs, the overall impression is decisive. The assessment must not be based on individual, isolated elements. Rather, a comprehensive evaluation of the signs to be compared is required. In particular, distinctive and dominant elements of the signs must be especially emphasized.[6]

The great importance of the distinctive and dominant elements of a sign is due to the fact that the average consumer cannot study the signs to be compared side by side, but rather at different times, for example, a few days apart. The distinctive and dominant elements are more likely to remain in the memory of the average consumer, so that the average consumer can only compare these defining elements of the signs.

8.3.2 Visual Comparison

The visual representation of a word mark is not relevant, as the individual letters of a word mark are protected. The spelling (font, upper or lower case, camel case, etc.) or color scheme can be arbitrary, as long as the spelling or color scheme is within the usual commercial practice.

The use of the same color or colors in a figurative mark or word/device mark can indicate visual similarity. However, the overall impression is decisive, and the same color scheme alone cannot establish similarity of signs.

8.3.3 Phonetic Comparison

For word marks, the phonetic comparison takes precedence over the visual comparison. In particular, a different spelling can still result in the word marks being pronounced in the same way. For example, individual letters can be replaced by other letters or a sequence of letters, without affecting the phonetic similarity. In the German 6 language, the following applies: The letters "P" and "B", "K" and "C" are pronounced similarly. The letter "V" can be replaced by the sequence "VAU" without any phonetic difference. In addition, "EX" could be replaced by "ECS" without changing the sound.

Example

The two marks "VW" and "Vauweh" sound identical in the German language. Here there is phonetic identity of the signs, so that overall, identity of the signs is to be

[6] ECJ, 11.11.1997, C-251/95, Gewerblicher Rechtsschutz und Urheberrecht, 1998, pp. 387-390,— Leaping Wildcat.

assumed. It is irrelevant that the two marks are visually dissimilar. Conceptual similarity is also excluded. However, it is sufficient for identity of the signs to manifest in one respect to assume overall identity of the two marks.◀

Example

For the two marks "Porsche" and "Borsche", phonetic similarity of the signs is to be assumed.◀

A word mark is to be pronounced as it is practiced by the relevant public. If a word mark is written in an unfamiliar language, the word mark is to be compared phonetically as the relevant public would pronounce it. If it is assumed that the relevant public would pronounce a word mark in a foreign language in several ways, a phonetic comparison must be made for all variants.

Due to significant visual or conceptual differences, a likelihood of confusion may be denied despite phonetic similarity between two marks.

A phonetic comparison is excluded for two figurative marks. Exceptionally, a phonetic comparison may be possible if the graphic elements directly and immediately suggest a meaning that can be pronounced. This may be particularly relevant if such a meaning is suggested by intensive advertising.

8.3.4 Conceptual Comparison

Conceptual similarity is also referred to as associative similarity. The marks must be examined conceptually, i.e., it must be determined whether the respective meanings of the marks can lead to a likelihood of confusion. Conceptual similarity exists if the two marks are mentally associated with each other.

Example

A word mark "BlueHouse" can be confused with a figurative mark depicting a blue-painted house.◀

Conceptual similarity can arise in particular between a word mark and a figurative mark, or between a word mark and the graphic elements of a word/device mark. It is not necessary for the meanings of the marks to be understood by all members of the relevant public. It is sufficient if a majority of the relevant public interprets the meanings in such a way that conceptual similarity is to be assumed.

If the marks consist of several elements, it is not sufficient if only individual components establish conceptual similarity between the marks. In this case, the value of the individual elements for the respective mark must be assessed in order to determine the significance of conceptual similarity of individual elements. It should be noted that the relevant public does not take an analytical approach, but usually focuses on the defining elements. This is all the more true since a direct comparison of the marks is not possible, but the marks are perceived sequentially. An analytical comparison of the marks is therefore generally excluded.

Conceptual meanings must be assessed in the context of the mark and the goods and services for which the mark is registered. If the meanings are closely related to the goods and services of the mark, they are of little significance. Meanings that cannot be associated with the goods and services are of greater significance.

8.3.5 Length of Signs

For word marks, the length of the signs is of great importance for visual perception. If the signs to be compared are very different in length, this argues against similarity of the signs.

For short signs, the individual letters of a word mark are perceived more precisely. Even small deviations in short signs result in a different overall impression and lead to a denial of likelihood of confusion.

8.3.6 Beginning and End of Words

The beginning of a word is more decisive for the overall impression of a sign than its ending. The ending is more important than the middle. A general rule is therefore that the beginning of a word is the most important part of a sign and should be given special consideration when assessing similarity of signs. However, this principle can be deviated from if, for example, the middle of the word is fanciful and the beginning and end are more descriptive for the goods and services for which the mark is registered.

Different endings of two signs to be compared are generally not sufficient to exclude similarity of signs. In individual cases, similarity of signs may be affirmed despite different beginnings.

8.3.7 Vowel Sequences

Vowels are more important than consonants. Vowels are so-called sonorants, i.e., a, e, i, o, and u. All other letters are referred to as consonants. If two marks have the same sequence of vowels, with different consonants in between, there may be similarity of signs.

Example

For the word mark “Pasqale”, the vowel sequence is “a-a-e”. A mark with the same vowel sequence can lead to similarity of the marks. For example, with the mark “Patmate”, which has the same vowel sequence but all consonants are different, there is more likely to be similarity of signs than with the mark “Pisqeli”, which has the same consonant sequence but different vowels.◀

8.3.8 Syllable Sequence

A syllable is a unit of consecutive sounds that can be pronounced in one go. A syllable is therefore a unit of speech. In the phonetic comparison of two signs, the number and sequence of syllables is of great importance for the overall phonetic impression. An identical syllable sequence in two signs is highly likely to result in phonetic similarity.

A similar number of syllables in two marks can also lead to similarity of signs. If the signs to be compared have a very different number of syllables, phonetic similarity is not to be assumed.

A general rule is therefore that a sign with a very small number of syllables and a sign with a large number of syllables are unlikely to be similar. However, it must be considered whether parts of the long mark should be disregarded, for example, if they are purely descriptive.

Example

The two marks “Mit Pasquale-Software Technik neu erleben” and “Pasquale”, each for software and computers, have a very different number of syllables. However, in the slogan mark, only the element “Pasquale” is distinctive. The remaining elements are descriptive for the goods software and computers and are therefore irrelevant. The two marks therefore have similarity of signs despite the different number of syllables.◀

A simple transposition of syllables cannot exclude similarity of signs.

Example

Best Software GmbH registers the mark “Pre-Care” for software. Bad Software GmbH sells computers three months later under the mark “Care-P.R.E”. Best Software GmbH can prohibit Bad Software GmbH from using the mark “Care-P.R.E” due to likelihood of confusion with its mark.◀

However, a transposition of syllables can lead to dissimilarity of two signs if it is an unusual syllable arrangement or if both are fanciful terms.

Example

Best Software GmbH names its software "Aronamics". Bad Software GmbH uses the mark "Icsaronam", where Bad Software GmbH has placed the last syllable of Best Software GmbH's mark at the beginning. The marks "Aronamics" and "Icsaronam" are dissimilar despite a simple transposition of syllables.◄

8.3.9 Three-Letter Marks

As a rule, for very short marks, especially marks with only three letters, the difference of just one letter is sufficient to deny similarity of signs. Such three-letter marks are usually acronyms. Acronyms are words whose letters are the initial components of several words.

Example

The two marks "Mer" and "Uer" are not similar. For the marks "Mer" and "Ner", the situation is to be assessed differently.◄

8.4 Multi-part Word Marks

For a multi-part word mark, the first step is to assess whether all components contribute to the distinctive character of the mark.[7] Descriptive elements of a word mark cannot be considered distinctive and are disregarded when assessing the similarity of signs. If the distinctive elements of two signs are identical or similar, similarity of signs is assumed.

8.5 Distinctiveness of the Earlier Mark

Patent offices distinguish between "strong" and "weak" marks. A "weak" mark is inherently attributed with low distinctiveness. A mark whose distinctiveness is not inherently weak, but which is not particularly well-known among the relevant public, is considered to have average distinctiveness.

[7] ECJ, 6.10.2005, C-120/04, Industrial Property Protection and Copyright Law, 2005, p. 1042—Thomson Life.

A "strong" mark can arise from recognition among the relevant public. A "strong" mark can also develop from a "weak" mark, i.e., a mark that, for example, is not very imaginative or is descriptive to a limited extent for the goods and services for which it is registered. There are also well-known or famous marks that are known to everyone. The greater the distinctiveness of the mark, the broader the scope of protection granted to it. Therefore, for a "strong" mark, there is generally a greater likelihood of confusion compared to a "weak" or "average" mark.

If an earlier mark is highly imaginative and has been extensively used, it is granted a broad scope of protection. This means that in such cases, similarity of signs is more likely to be assumed.

Example

A mark such as "Pasquale" or "Aronautics" for software is granted a broader scope of protection than a mark like "BestSoft" for software.◄

In the absence of special circumstances, an earlier mark is assumed to have average recognition. If the earlier mark is a famous mark, it is assumed to have a broad scope of protection, which means that a likelihood of confusion is more likely. A famous mark, in particular, extends its influence to other classes.

Example

The "Davidoff" mark for cigars is so well known that even for dissimilar goods, such as belts, a likelihood of confusion is assumed if there is similarity of signs.[8]◄

On the other hand, an earlier mark may have very low distinctiveness if it is descriptive for the goods and services.

8.6 Interaction Between Similarity of Signs and Similarity of Goods

Low similarity of signs can be compensated by a high similarity of goods and services, so that a likelihood of confusion may still be assumed. The same applies to low similarity of goods and services and high similarity of signs. Conversely, if there is no similarity of signs or no similarity of goods and services, a likelihood of confusion is excluded.

[8] ECJ, 9.1.2003, C-292/00, Industrial Property Protection and Copyright Law, 2003, p. 240—Davidoff.

8.7 Relevant Public

For marks aimed at a highly attentive public, a higher degree of similarity between the signs and goods is required for a likelihood of confusion to be assumed.

Example

In the case of pharmaceuticals, the public pays very close attention to differences. Therefore, for goods such as pharmaceuticals, two marks must be very similar to establish a likelihood of confusion. If the relevant public consists of businesspeople rather than consumers, a high level of attention is also assumed. A counterexample is frozen pizzas in supermarkets for consumers. In this case, the relevant public is generally not particularly discerning.◄

For the relevant public, the assumption is initially that of an average, reasonably well-informed, observant, and circumspect consumer. If the relevant public consists of business customers, the assessment of the likelihood of confusion is made from the perspective of an average, reasonably well-informed, observant, and circumspect business customer.

The level of attention of the public is generally lower for everyday goods and services compared to the attention paid to technical or highly specialized goods and services, such as pharmaceuticals or special-purpose machinery.

If the relevant public typically possesses professional knowledge, an above-average level of expertise is assumed for the average consumer or average business customer.

For expensive or rare goods, it is assumed that average consumers exhibit a very high level of attention. Examples of expensive goods include cars and condominiums. The purchase of life insurance can be considered rare.

In the case of risky purchases, an above-average level of attention can also be assumed. An example of this is the purchase of a prescription medication.

8.8 Famous Marks

There are certain famous marks for which a simplified assessment of the likelihood of confusion is applied, as the fame of the mark is so great that its recognition extends to all classes. For a mark that is famous domestically, it is therefore not necessary to assess the similarity of goods and services. For a likelihood of confusion, similarity of signs alone is sufficient.[9] An example of such a famous mark is the "Coca-Cola" script of the Coca-Cola Company.

[9] Section 14(2) No. 3 German Trademark Act.

Research 9

A search for prior trademark rights can be conducted using the databases of the patent offices. The use of these databases is free of charge. The databases are generally available online.

9.1 Vienna Classification

The Vienna Classification is used to make figurative marks and figurative elements of word/figurative marks searchable. For this purpose, a hierarchically structured system was developed, branching from the general to the specific. The sequence of branches for a specific image is represented by a so-called notation. A notation is a sequence of numbers. With this notation, figurative marks or figurative elements of word/figurative marks can be checked for similarity.

Examples of notation according to the Vienna Classification:[1]

- "Cheese in round form" is assigned to category 8 (foodstuffs) and has the notation 8.3.9
- "Trousers" are assigned to category 9 (textiles, clothing, sewing utensils, headgear, footwear) and have the notation 9.3.3
- "Watches" are assigned to category 17 (watches, jewelry, measures and weights) and have the notation 17.1.1

[1] DPMA, "https://www.dpma.de/docs/marken/markenwienerklassifikation_8-2018.pdf", accessed on June 11, 2021.

T. H. Meitinger, *Trademark Without a Lawyer*,
https://doi.org/10.1007/978-3-662-73127-7_9

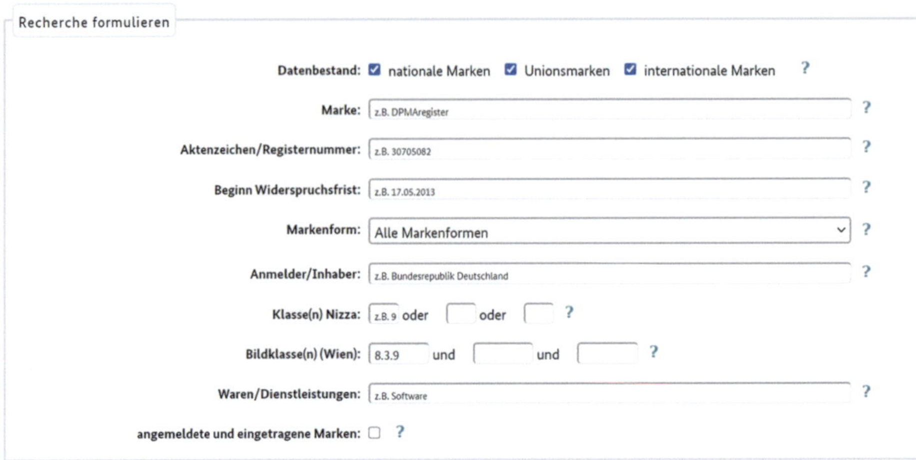

Fig. 9.1 Search using notation according to Vienna Classification (DPMA)

Figure 9.1 shows the search mask of the DPMARegister[2] in which up to three notations can be entered under "Image class(es) (Vienna)".

9.2 Register of the German Patent Office

The register of the German Patent Office is available to everyone for research purposes (see Fig. 9.2).[3]

In the input mask of the German Patent Office, the trademark, the type of mark (in particular word mark, word/figurative mark, or figurative mark), and the Nice classes (maximum of three classes) can be entered. In addition, the goods and services can be specified. The input mask also allows the entry of the Vienna class to search for figurative elements of word/figurative marks or figurative marks.

A disadvantage of searching in the register of the German Patent Office is that it is not possible to search for similar marks. Of course, it is possible to truncate the entries to search for similar marks. However, this results in a high degree of uncertainty.

[2] DPMA, "https://register.dpma.de/DPMAregister/marke/basis", accessed on June 11, 2021.

[3] DPMA, "https://register.dpma.de/DPMAregister/marke/basis", accessed on June 16, 2021.

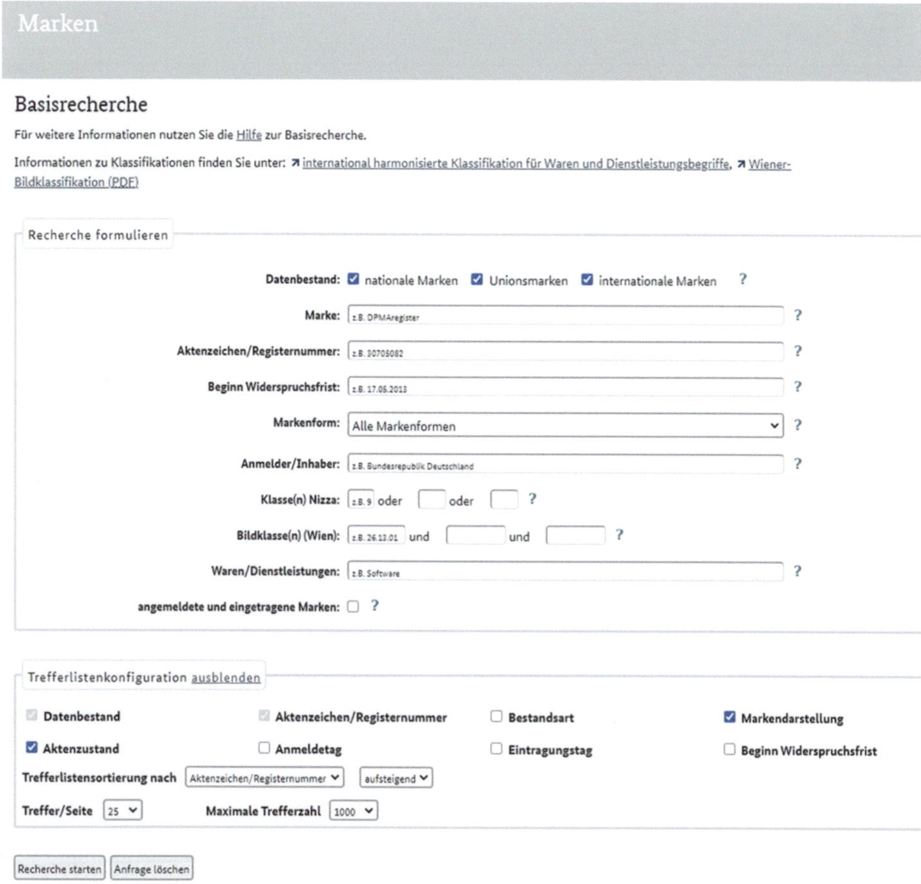

Fig. 9.2 Basic search (DPMA)

9.3 TMview of the EUIPO

With the TMview database[4] it is possible to search for similar marks (see Fig. 9.3).[5]

This database is operated by the EUIPO[6]. When using the EUIPO's online database, it is essential to restrict the goods and services, in particular by specifying the relevant Nice classes. Otherwise, a large number of irrelevant search results may be generated.

---

[4] TMView: TM for Trademark and View for viewing.

[5] EUIPO, "https://www.tmdn.org/tmview/#/tmview", accessed on June 2, 2021.

[6] EUIPO is the abbreviation for European Union Intellectual Property Office, based in Alicante (Spain), until March 23, 2016: Office for Harmonization in the Internal Market (Trade Marks and Designs).

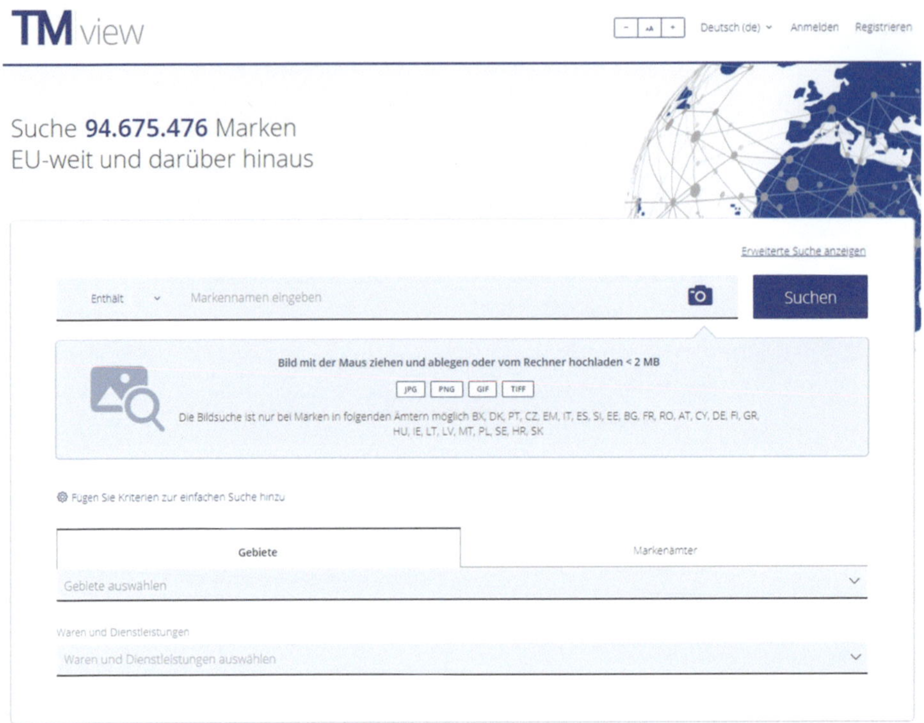

Fig. 9.3 TMview (EUIPO)

The advantage of the EUIPO database is that searches can be conducted for numerous countries. However, it is not possible to search for trademarks from one important country in the EUIPO database, namely Swiss trademarks.

Fortunately, the TMview database allows for a similarity search without the need for truncation. Therefore, with a high degree of certainty, all marks that could lead to a likelihood of confusion with the entered mark are displayed.

9.4 Global Brand Database of WIPO

The Global Brand Database is the online database of WIPO.[7]

The WIPO input mask is not available in German. The input mask features the tabs Brand, Names, Numbers, Dates, Class, and Country. Under "Brand", the word mark or the text component of a word/figurative mark can be entered in the "Text" field. The

[7] WIPO, "https://www3.wipo.int/branddb/en/", accessed on June 2, 2021.

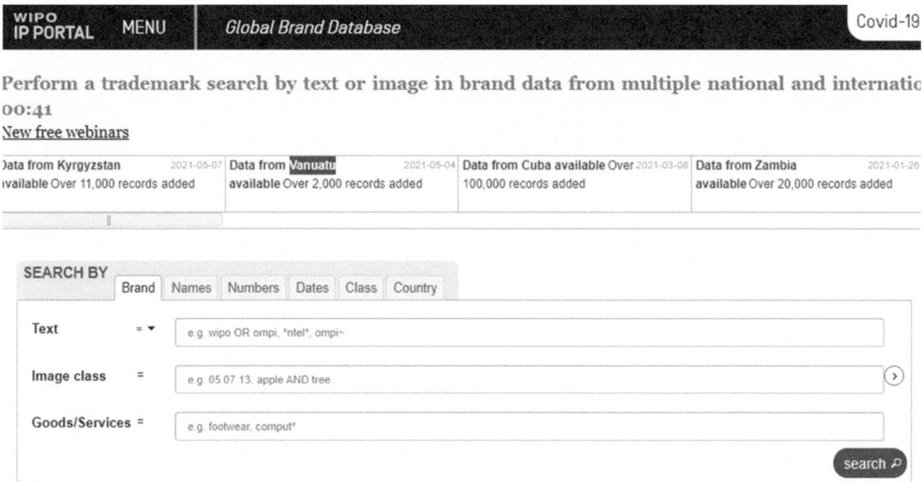

Fig. 9.4 Global Brand Database (WIPO)

Vienna class (Image class) can also be entered to search for the figurative elements of word/figurative marks or for figurative marks (see Fig. 9.4).

Under the "Names" tab, the applicant (Holder) can be entered in the input mask. The great advantage of the WIPO database is that "Switzerland" can also be entered under the "Country" tab. Therefore, Swiss trademarks can also be searched for (see Fig. 9.5).

9.5 Google Alert

Google Alert can be used to monitor use-based marks. Google Alert can be set up at the link "https://www.google.de/alerts".

9.6 German Trademark versus European Union Trademark

An opposition can be filed against a German trademark with an earlier German trademark, a European Union trademark, and with an international registration designating Germany or EM[8]. A nullity procedure before the patent office or a cancellation action before a regular court can also be initiated. Therefore, these prior rights must be searched if a German trademark is to be registered.

A European Union trademark can be challenged with a European Union trademark, a national trademark of an EU Member State, or an international registration designating

[8] EM is the abbreviation for a European Union trademark in an international registration.

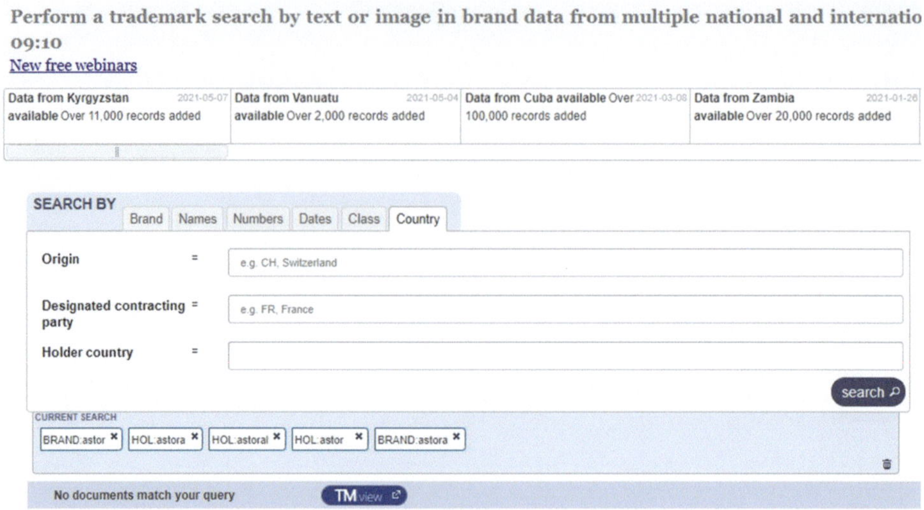

Fig. 9.5 Global Brand Database—Names—(WIPO)

EM or an EU Member State. Therefore, when applying for a European Union trademark, searches must be conducted for all EU countries and the corresponding international registrations.

German Trademark 10

With a registered German trademark, trademark protection is acquired within the territory of Germany. The prerequisite is the entry of the trademark into the trademark register of the German Patent Office. Before registration, the German Patent Office examines the so-called absolute grounds for refusal[1]. The Patent Office does not examine whether the trademark poses a likelihood of confusion with an earlier trademark.[2]

10.1 Deadlines for a German Trademark

The submission of the application documents triggers the start of the deadlines for the trademark. For example, the application fee for the trademark registration must be paid within three months of the filing date. In Table 10.1, the deadlines for a filing date of July 10, 2021 are calculated as an example.

A trademark can be renewed any number of times, each for a period of 10 years. The renewal fee, including up to three classes, is € 750.

Table 10.2 shows, as an example, the deadlines for payment of renewal fees for a filing date of July 10, 2021. The renewal fees are due six months before the expiry of the respective protection period and can be paid without surcharge until the end of the protection period.

[1] See Chap. 7: Grounds for refusal.

[2] See Chap. 8: Likelihood of confusion.

T. H. Meitinger, *Trademark Without a Lawyer*,
https://doi.org/10.1007/978-3-662-73127-7_10

Table 10.1 Deadlines for a German trademark

Deadlines for a German Trademark	
Filing date	July 10, 2021
Payment of application fee (within 3 months after filing date)	October 10, 2021
Priority period (6 months after filing date)	January 10, 2022
Grace period for non-use (5 years after registration or after the end of an opposition proceeding)	after July 10, 2026

Table 10.2 Deadlines for payment of renewal fees

Deadlines for renewal fees		
1	Renewal fee	January 10, 2031
2	Renewal fee	January 10, 2041
3	Renewal fee	January 10, 2051
4	Renewal fee	January 10, 2061
5	Renewal fee	January 10, 2071
6	Renewal fee	January 10, 2081
7	Renewal fee	January 10, 2091
8	Renewal fee	January 10, 2101

Fig. 10.1 Fees for a German trademark (as of October 2021)

Fees for a German trademark	
• Basic fee for trademark registration	
– in paper form (including 3 classes)	300 euros
– electronically (including 3 classes)	290 euros
• Each additional class	100 euros

10.2 Fees for a German Trademark

Registering a German trademark by mail costs € 300. This amount includes up to three classes.

If the trademark is submitted electronically to the Patent Office, the trademark application costs €290. This amount includes up to three classes. Each additional class costs an extra € 100 (see Fig. 10.1). The current fees can be found in the fee schedule of the German Patent Office at the link "https://www.dpma.de/docs/formulare/allgemein/a9510.pdf". Online application is possible via the DPMAdirektWeb web portal[3].

[3] DPMA, "https://direkt.dpma.de/marke/", accessed on June 21, 2021.

10.3 Priority

A German trademark can claim the priority of a foreign trademark. By validly claiming priority, the German trademark is assigned the earlier priority date of the foreign trademark. The prerequisite is that the German trademark is filed with the German Patent Office within six months of the filing date of the foreign trademark.[4]

Only the priority of a foreign trademark can be claimed, provided that the trademark representation and the goods and services are identical to those of the German trademark. The priority of an earlier German trademark cannot be claimed by a German trademark application.[5]

10.4 Search for Earlier Rights

An opposition, a cancellation procedure before the German Patent Office, or a cancellation action before a civil court against a German trademark can only be based on earlier German trademarks, EU trademarks, and international registrations designating Germany and EM[6]. Therefore, when searching for earlier rights, only German trademarks, EU trademarks, and international registrations designating Germany and EM should be considered. For international registrations, it should not be overlooked that Germany or EM can be subsequently designated.

10.5 Withdrawal and Surrender

The trademark owner can completely withdraw their trademark or surrender it for individual goods and services of their trademark.[7] This option of withdrawal or partial surrender of the trademark can be advantageous if the owner of an earlier trademark asserts a claim for injunctive relief against the trademark owner. By surrendering individual goods and services, it may be possible to eliminate the likelihood of confusion or at least persuade the owner of the earlier trademark to accept coexistence of the two trademarks.

---

[4] Article 4 C paragraph 1 Paris Convention for the Protection of Industrial Property (PCPIP).

[5] In patent law, an "internal" priority is possible. Here, a German patent application can claim the priority of an earlier German patent application (§ 40 paragraph 1 Patent Act). In trademark law, there is no internal priority.

[6] EM is the abbreviation for an EU trademark in the context of an international registration.

[7] Section 39 paragraph 1 Trademark Act and Section 48 paragraph 1 Trademark Act.

By withdrawing a trademark, an opposition or cancellation procedure against this trademark can be terminated. Withdrawal is possible at any time and does not require the consent of the opposing party. A withdrawal or surrender, once made, cannot be revoked.[8]

[8] Federal Court of Justice (BGH), March 14, 1985, X ZB 13/83, Gewerblicher Rechtsschutz und Urheberrecht, 1985, p. 919—Caprolactam; BGH, December 7, 1976, X ZB 24/75, Gewerblicher Rechtsschutz und Urheberrecht, 1977, p. 485—Withdrawal of the patent application.

11 EU Trademark

A European trademark is called a European Union trademark.[1] A European Union trademark provides protection in all EU member states. The EU currently has 27 member states (see Table 11.1).[2]

11.1 Deadlines for a European Union Trademark

The following Table 11.2 provides sample deadlines for a European Union trademark with a filing date of June 16, 2021. The application fee must be paid no later than one month after the filing date. The so-called grace period for non-use ends five years after the trademark is entered in the EUIPO register.

The following Table 11.3 provides sample due dates for payment of renewal fees for a filing date of June 16, 2021. Renewal fees can be paid from these respective dates until the end of the protection period without surcharge.

11.2 Fees for a European Union Trademark

The official application fee is € 850. Only one class is included. A second class costs € 50. Each additional class costs an extra € 150 (see Fig. 11.1).

[1] Until March 2016, a European trademark was referred to as a Community trademark.

[2] Until 2020, the United Kingdom was a member state of the EU, and a European Union trademark therefore also provided protection in that country.

T. H. Meitinger, *Trademark Without a Lawyer*,
https://doi.org/10.1007/978-3-662-73127-7_11

Table 11.1 Trademark protection through European Union trademark

Member States of the European Union (EU)	
Belgium	Malta
Bulgaria	The Netherlands
Denmark	Austria
Germany	Poland
Estonia	Portugal
Finland	Romania
France	Sweden
Greece	Slovakia
Ireland	Slovenia
Italy	Spain
Croatia	The Czech Republic
Latvia	Hungary
Lithuania	Cyprus
Luxembourg	

Table 11.2 Deadlines for a European Union trademark

Deadlines for a European Union trademark	
Filing date	June 16, 2021
Payment of application fee (within one month of filing date)	July 16, 2021
Priority period (6 months after filing date)	December 16, 2021
Grace period for non-use (5 years after registration)	after June 16, 2026

Table 11.3 Deadlines for renewal fees of a European Union trademark

Deadlines for renewal fees		
1	Renewal fee (6 months before the end of the 10-year period)	December 16, 2030
2	Renewal fee (6 months before the end of the 10-year period)	December 16, 2040
3	Renewal fee (6 months before the end of the 10-year period)	December 16, 2050
4	Renewal fee (6 months before the end of the 10-year period)	December 16, 2060
5	Renewal fee (6 months before the end of the 10-year period)	December 16, 2070
6	Renewal fee (6 months before the end of the 10-year period)	December 16, 2080
7	Renewal fee (6 months before the end of the 10-year period)	December 16, 2090
8	Renewal fee (6 months before the end of the 10-year period)	December 16, 2100

Fig. 11.1 Fees for a European Union trademark (as of October 2021)

Fees for a European Union trademark	
• Basic fee for trademark application	
– electronically (including 1 class)	850 euros
• 2nd grade	50 euros
• 3rd grade and each additional grade	150 euros

A European Union trademark is approximately three times as expensive as a German trademark. However, considering that a European Union trademark provides trademark protection in 27 countries, the fees for a European Union trademark appear reasonable.

11.3 Priority

A European Union trademark can claim the priority of a national trademark. The prerequisite is that the filing date of the national trademark is no more than six months before the filing date of the European Union trademark.[3] In addition, the trademark establishing priority must be of the same type and representation as the European Union trademark. The goods and services of the European Union trademark must be the same as those of the trademark establishing priority, or must be covered by the goods and services of the trademark establishing priority.[4] Claiming priority has the effect that the European Union trademark assumes the earlier priority date of the national trademark establishing priority.

11.4 Seniority

After the priority period has expired, an earlier priority date of a national trademark can still be transferred to a European Union trademark through seniority.[5] However, the earlier priority date only applies to the respective national portion of the European Union trademark, and only the seniority of a national trademark of an EU member state can be used.[6]

The legal concept of seniority can be used to streamline a trademark portfolio, as national trademarks can be abandoned after their earlier priority date has been transferred to the European Union trademark. This results in a leaner trademark portfolio and reduced costs.

[3] Article 4 C paragraph 1 Paris Convention, "https://www.wipo.int/edocs/pubdocs/de/intproperty/201/wipo_pub_201.pdf", accessed June 12, 2021.

[4] Article 34 paragraph 1 European Union Trademark Regulation.

[5] Article 39 paragraph 1 and Article 40 paragraph 1 European Union Trademark Regulation.

[6] General Court, 19.6.2014, T-382/12, BeckRS, 2014, 81.630—Nobel/Nobel.

A disadvantage of seniority is that a potential challenger only needs to attack the European Union trademark to invalidate trademark protection as a whole. An attack on a European Union trademark is also more likely, since any owner of a national trademark in an EU member state can challenge a European Union trademark. A national trademark of an EU member state can only be challenged by older national trademarks from the same country, by European Union trademarks, and by international registrations designating the respective country or EU trademark[7].

11.5 Observations by Third Parties

Any natural or legal person may submit so-called "observations by third parties" regarding a pending trademark application at the EUIPO, stating reasons that may prevent the registration of the trademark. Only reasons concerning absolute grounds for refusal are considered by the office.[8] In particular, the observations may be based on lack of distinctiveness, infringement of the need to keep certain signs free, or the risk of deception among the relevant public.[9] Observations by third parties are included in the examination procedure by the office until the end of the opposition period or, if opposition proceedings are pending, until a decision is made on the opposition.[10] Earlier rights are not considered as observations by third parties. The opposition procedure is provided for asserting earlier rights.[11]

11.6 Search Report

The EUIPO prepares a search report for every application, listing earlier European Union trademarks that may be confusingly similar to the applied-for trademark.[12] This search report is sent to the owners of the earlier European Union trademarks after publication of the application.[13] This enables the owners of the earlier European Union trademarks to file an opposition against the registration of a later trademark at the EUIPO.[14]

[7] EU trademark (EM) is the abbreviation for a European Union trademark in an international registration.

[8] Article 45 paragraph 1 European Union Trademark Regulation.

[9] Article 7 European Union Trademark Regulation.

[10] Article 45 paragraph 2 European Union Trademark Regulation.

[11] Article 46 European Union Trademark Regulation.

[12] Article 8 European Union Trademark Regulation.

[13] Article 43 paragraph 7 sentence 1 European Union Trademark Regulation.

[14] Article 46 European Union Trademark Regulation.

The applicant of the later trademark may request that the search report on earlier rights that may conflict with their European Union trademark be sent to them.[15] No additional costs are incurred by the applicant for this.

11.7 Waiver

The owner of a trademark may waive their entire trademark or individual goods and services.[16] By waiving their trademark, the trademark owner can terminate opposition or cancellation proceedings before the EUIPO against their trademark. It is not necessary for the opposing party to consent to the waiver. A waiver must be declared in writing.[17] By waiving individual goods and services, it may be possible to eliminate a likelihood of confusion with an earlier trademark in the event of a conflict.

11.8 German Trademark or European Union Trademark?

The question of the required territorial scope of trademark protection can only be answered on a case-by-case basis. For example, if goods and services are to be offered only in Germany, a German trademark appears sufficient. However, if it is also necessary to ensure that the trademark is not used by third parties in other European countries, a European Union trademark is required. When applying for a European Union trademark, a similarity search should be conducted for all EU member states.

[15] Article 43 paragraph 6 European Union Trademark Regulation.

[16] Article 57 paragraph 1 European Union Trademark Regulation.

[17] Article 57 paragraph 2 sentence 1 European Union Trademark Regulation.

International Registration 12

With an international registration at WIPO in Geneva, trademark protection can be obtained in most countries worldwide. The applicant must pay a basic fee to WIPO and a respective national fee for each desired country. The applicant can extend their trademark protection to any countries of their choice. In this way, for example, a German trademark can be extended to the USA.

A European Union trademark can be obtained through international registration. For example, a German trademark can serve as the basic mark and be extended to the EU via international registration. Advantageously, international registration can also provide trademark protection in the USA, China, etc.

A prerequisite for international registration is a so-called basic mark. A basic mark can be a national or a regional trademark application. With an international trademark application, international trademark protection can be achieved cost-effectively.

12.1 Madrid Agreement Concerning the International Registration of Marks

The legal basis for international registration is the Madrid Agreement Concerning the International Registration of Marks (Madrid Agreement, MMA). The Madrid Agreement dates back to 1891. Major industrialized nations agreed in the treaty that a single trademark registration could be extended to all countries that are contracting parties. In 1989, an extension was adopted, namely the Protocol Relating to the Madrid Agreement, through which additional countries joined the Madrid Union. The Madrid Agreement and its Protocol together have 106 member states.

T. H. Meitinger, *Trademark Without a Lawyer*,
https://doi.org/10.1007/978-3-662-73127-7_12

The procedure for international registration is that a basic mark must first be applied for. For example, a national German trademark or a European Union trademark can serve as the basic mark. The application documents can be filed with the German Patent Office. These are forwarded to WIPO within two months. WIPO then transmits the trademark application to the national patent offices designated by the applicant. This results in a bundle of national trademark applications.

A prerequisite for international registration is that the applicant has a residence or a place of business in one of the member states of the Madrid Agreement or its Protocol. A basic fee must be paid, as well as a fee that depends on the designated countries. In addition, a fee must be paid corresponding to the number of claimed Nice classes.

After WIPO has forwarded the application documents to the individual national countries, the national authorities examine the trademark applications according to their respective national trademark laws. The respective national trademark law can differ significantly from German trademark law. For example, the USPTO[1] also examines for prior rights before registration.

Subsequent designation is possible, meaning that trademark rights can be subsequently applied for in additional states party to the Madrid Agreement. A fee is required for this.

When registering internationally, it should be taken into account that each country may have specific designations for goods and services. Fortunately, the EUIPO has provided in its TMclass database[2] a feature that allows goods and services to be specified according to country.

Fifty-five states have acceded to the Madrid Agreement. All member states of the Madrid Agreement have also joined the Protocol to the Madrid Agreement. Therefore, all member states of the Madrid Agreement are full members.

12.2 Protocol Relating to the Madrid Agreement

In 1989, the Protocol Relating to the Madrid Agreement (Protocol to the Madrid Agreement Concerning the International Registration of Marks, PMMA) was adopted. The PMMA is legally independent of the MMA. All member states of the MMA have joined the PMMA. The collective group of states that have joined either the MMA or the PMMA is referred to as the Madrid Union. The Madrid Union comprises 106 states (see Table 12.1).

[1] USPTO United States Patent and Trademark Office.

[2] TMclass: TM for Trademark and class for class.

Table 12.1 Member States of the MMA and PMMA

PMMA Member States	MMA Member States
Afghanistan	
African Intellectual Property Organization (OAPI)	
Albania	x
Algeria	x
Antigua and Barbuda	
Armenia	x
Australia	
Austria	x
Azerbaijan	x
Bahrain	
Belarus	x
Belgium	x
Bhutan	x
Bosnia and Herzegovina	x
Botswana	
Brazil	
Brunei Darussalam	
Bulgaria	x
Cambodia	
Canada	
China	x
Colombia	
Croatia	x
Cuba	x
Cyprus	x
Czech Republic	x
Democratic People's Republic of Korea	x
Denmark	
Egypt	x
Estonia	
Eswatini	
European Union (EU)	

(continued)

Table 12.1 (continued)

PMMA Member States	MMA Member States
Finland	
France	x
Gambia	
Georgia	
Germany	x
Ghana	
Greece	
Hungary	x
Iceland	
India	
Indonesia	
Iran (Islamic Republic of)	x
Ireland	
Israel	
Italy	x
Japan	
Kazakhstan	x
Kenya	x
Kyrgyzstan	x
Lao People's Democratic Republic	
Latvia	x
Lesotho	x
Liberia	x
Liechtenstein	x
Lithuania	
Luxembourg	x
Madagascar	
Malawi	
Malaysia	
Mexico	
Monaco	x
Mongolia	x

(continued)

Table 12.1 (continued)

PMMA Member States	MMA Member States
Montenegro	x
Morocco	x
Mozambique	x
Namibia	x
Netherlands	x
New Zealand	
North Macedonia	
Norway	
Oman	
Pakistan	
Philippines	
Poland	x
Portugal	x
Republic of Korea	
Republic of Moldova	x
Romania	x
Russian Federation	x
Rwanda	
Samoa	
San Marino	x
Sao Tome and Principe	
Serbia	x
Sierra Leone	x
Singapore	
Slovakia	x
Slovenia	x
Spain	x
Sudan	x
Sweden	
Switzerland	x
Syrian Arab Republic	x
Tajikistan	x

(continued)

Table 12.1 (continued)

PMMA Member States	MMA Member States
Thailand	
Trinidad and Tobago	
Tunisia	
Turkey	
Turkmenistan	
Ukraine	x
United Kingdom	
United States of America	
Uzbekistan	
Vietnam	x
Zambia	
Zimbabwe	

12.3 EU Trademark or German Trademark as Basic Mark?

An international registration can be applied for using either a German trademark or a European Union trademark as the basic mark. Germany is a full member, as it has signed both the Madrid Agreement and the Protocol to the Madrid Agreement. The EU (European Union) is only a protocol member, as it has only joined the Protocol to the Madrid Agreement. Other protocol members include, for example, the USA and Japan. For an international registration with a basic mark from a protocol member, some full members charge significantly higher fees compared to an international application with a basic mark from a full member.

An international registration is dependent on the basic mark for five years. This means that within this period, the cancellation of the basic mark will result in the loss of the international registration and all national or regional trademarks derived from it. A German trademark can only be challenged by marks that are effective in Germany. The most common reason for cancellation is opposition by a third party.

For a European Union trademark, attacks can be made from any EU member state. A European Union trademark can be challenged by a German trademark, an Italian trademark, a Greek trademark, a French trademark, etc. It is therefore more likely that a European Union trademark will be attacked by an opposition. This is especially true since the EUIPO itself conducts a similarity search and notifies holders of earlier rights about potentially confusingly similar, later trademarks.

Table 12.2 Deadlines for an IR Mark

Deadlines for an IR Mark	
Filing date	June 16, 2021
Payment of fees (within one month after filing date)	July 16, 2021
first renewal of protection period	December 16, 2030
second renewal of protection period	December 16, 2040

It is therefore advisable to use a German trademark, rather than a European Union trademark, as the basic mark for an international registration. A European Union trademark can then be obtained via international registration. This approach results in higher costs, as a German trademark must be paid for in addition, which could also be obtained via international registration; however, this is the less risky route.

12.4 Subsequent Designation of Countries

Additional countries can be subsequently designated for an existing internationally registered trademark. For this, an application for subsequent designation must be filed. The application can be accessed at the link "https://www.wipo.int/export/sites/www/madrid/en/forms/docs/form_mm4.pdf".[3] The application can be filed with the German Patent Office or directly with WIPO.

12.5 Deadlines for an IR Mark

Table 12.2 lists the deadlines for an international registration, using June 16, 2021 as an example filing date.

The fees must be paid within the first month after the filing date. The international registration can be renewed any number of times for ten years each by paying a renewal fee. The renewal fee must be paid within a six-month period before the expiration of the protection period without surcharge.

[3] WIPO, Application MM4, "https://www.wipo.int/export/sites/www/madrid/en/forms/docs/form_mm4.pdf", accessed on June 11, 2021.

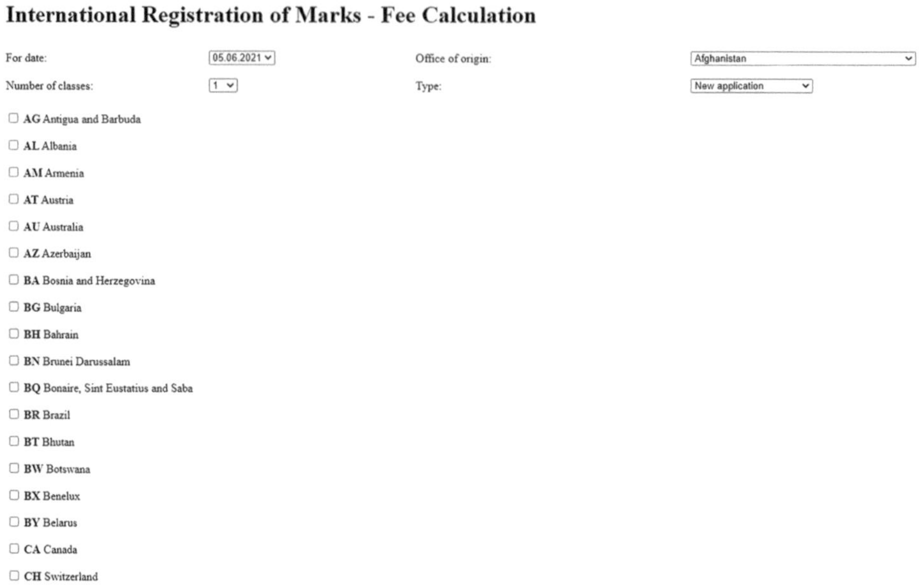

Fig. 12.1 Costs of an IR Mark (WIPO)

12.6 Fees for an IR Mark

The WIPO fee calculator can be accessed at the link “https://www.wipo.int/madrid/fee calc/FirstStep”.[4] In the fee calculator, the country of the basic mark can be entered (office of origin) and the number of desired classes (number of classes). If a mark from a country that is a full member is chosen as the basic mark, the costs are lower compared to a country that is only a protocol member. In addition, the countries to which trademark protection is to be extended can be selected (by clicking “continue,” the country list opens) (see Fig. 12.1).

[4] WIPO, “https://www.wipo.int/madrid/feecalc/FirstStep”, accessed on June 21, 2021.

13 Registering a Trademark

The required information for an application is discussed. A key focus is the compilation of goods and services. In addition, the different application methods—by mail, fax, or online—are presented.

13.1 Information about the Applicant

The name and address of the applicant must be provided correctly. The correct name and address of a GmbH or UG are those recorded in the commercial register. Before a trademark is registered for a GmbH or UG, it should be checked whether the company details match those in the commercial register.

For a civil law partnership (GbR, Gesellschaft bürgerlichen Rechts), it is necessary to provide at least the details of one partner. The reason for this is that a civil law partnership is not considered a legal entity like a GmbH or UG and is therefore not fully capable of holding rights and conducting business.

13.2 Goods and Services

The goods and services for which the trademark is to be registered must be specified precisely. It is not sufficient to simply indicate the individual classes. For each class, the terms that define the goods and services must also be provided. Merely stating the classes would be an indeterminate specification of goods and services, which would not lead to the registration of the trademark in the register.

T. H. Meitinger, *Trademark Without a Lawyer*,
https://doi.org/10.1007/978-3-662-73127-7_13

It is advisable to use the terms for defining goods and services as specified by the Nice Classification. These terms can be assumed to be sufficiently specific. Terms such as "machine" or "system" do not meet this criterion of specificity, for example.

13.2.1 Grouping Requirement

The goods and services must be grouped according to the Nice Classification. This means that for each Nice class, the desired goods and services must be specified. The selected Nice classes must be provided to the patent office in ascending numerical order.

13.2.2 Nice Classification

The designations of goods and services accepted by a patent office can be taken from the Nice Classification[1]. The Nice Classification was established on June 15, 1957, with the "Nice Agreement concerning the International Classification of Goods and Services for the Purposes of the Registration of Marks" by leading industrial companies. The Nice Classification is updated every five years. Currently, the eleventh edition from January 1, 2019, is valid, referred to as "Version 2021".[2]

The Nice Classification comprises 45 classes. All possible goods and services are assigned to classes. There are separate Nice classes for goods and specific classes for services. For example, the service "providing food and drink in restaurants" is listed in Nice class 43. The good software is included in Nice class 9 (see Table 13.1).

The EUIPO maintains the TMclass service[3]. At the link "http://euipo.europa.eu/ec2/?lang=de"[4] the Nice classes for goods and services can be queried. In this online portal, individual goods or services can be entered into the search field, and an assignment to Nice classes will be suggested. Alternatively, a Nice class can be entered in this online portal. In this case, all goods and services of that Nice class are presented.

It is essential to determine exactly for which goods and services a trademark is to be registered. While free wording is possible, it is not recommended. Rather, it is advisable to use the designations according to the requirements of the patent offices. Otherwise, the registration of the trademark at the patent office cannot be carried out automatically,

[1] Nice Agreement concerning the international classification of goods and services for the purposes of the registration of marks (short designation: Nice Classification).

[2] The Nice trademark classifications can be downloaded from the German Patent Office at the link "https://www.dpma.de/marken/klassifikation/waren_dienstleistungen/nizza/index.html" (DPMA, accessed May 29, 2021).

[3] TMclass: TM for Trademark and class for class.

[4] EUIPO, "http://euipo.europa.eu/ec2/?lang=de", accessed May 29, 2021.

Table 13.1 Nice classes

Goods	
Class 1	Chemical products for industrial, scientific, photographic, agricultural, horticultural, and forestry purposes
Class 2	Paints, dyes, and anti-corrosion agents
Class 3	Non-medical products for personal and beauty care as well as cleaning agents
Class 4	Technical oils and greases, fuels, and lighting materials
Class 5	Pharmaceutical products and other preparations for medical or veterinary purposes
Class 6	Raw and partly processed base metals, including ores, as well as certain goods made of base metals
Class 7	Machines and machine tools, engines, and drives; The term 'machines' needs to be specified more precisely, as it is not sufficiently clear and unambiguous.
Class 8	Hand-operated tools and implements for performing various tasks, such as drilling, shaping, cutting, and piercing
Class 9	Apparatus and instruments for scientific or research purposes, audiovisual and information technology equipment, as well as safety and rescue equipment
Class 10	Surgical, medical, dental, and veterinary apparatus, instruments, and items generally used for the diagnosis, treatment, or improvement of the physical functions or health of humans and animals
Class 11	Apparatus and installations for regulating environmental conditions, especially for lighting, cooking, cooling, and hygiene purposes
Class 12	Vehicles and apparatus for transporting people or goods by land, air, or water
Class 13	Firearms and pyrotechnic products
Class 14	Precious metals and certain items made from or coated with them, as well as jewelry, ornaments, watches, and their components
Class 15	Musical instruments, their parts, and accessories
Class 16	Paper, cardboard, and certain goods made from these materials, as well as office supplies
Class 17	Materials for insulating electricity, heat, or sound, and plastics for use in manufacturing processes in the form of films, sheets, or rods, as well as certain goods made of rubber, gutta-percha, gum, asbestos, mica

(continued)

Table 13.1 (continued)

Class 18	Leather, imitations of leather, and goods made from these materials
Class 19	Non-metallic materials for construction purposes
Class 20	Furniture and furniture parts, as well as certain goods made of wood, cork, reed, cane, wicker, horn, bone, ivory, whalebone, shell, amber, mother-of-pearl, meerschaum, and their substitutes or of plastic
Class 21	Small, hand-operated household and kitchen devices, as well as cosmetic devices and devices for personal and beauty care, glassware, and certain goods made of porcelain, ceramics, stoneware, terracotta, or glass
Class 22	Canvas and other materials for making sails, ropes, upholstery and filling materials, and raw textile fibers
Class 23	Natural or synthetic yarns and threads for textile purposes
Class 24	Fabrics, blankets, and covers for household use
Class 25	Clothing, footwear, and headgear for people
Class 26	Haberdashery, trimmings, real or artificial hair, as well as hair ornaments and small decorative elements used to embellish objects
Class 27	Coverings and claddings for floors and walls
Class 28	Toys, games, sports equipment, entertainment items, novelties, and certain articles for Christmas trees
Class 29	Foods of animal origin as well as vegetables and other edible horticultural products prepared or preserved for consumption
Class 30	Prepared or preserved foods of plant origin, except fruit and vegetables, as well as additives for enhancing the flavor of foods
Class 31	Unprepared products from the soil and sea, live animals and plants, as well as animal feed
Class 32	Non-alcoholic beverages and beers
Class 33	Alcoholic beverages, essences, and extracts
Class 34	Tobacco and smokers' articles, as well as certain accessories and containers related to their use
Services	

(continued)

Table 13.1 (continued)

Class 35	Advertising; business management, organization, and administration; office functions
Class 36	Financial services, monetary affairs, and banking services; insurance services; real estate affairs
Class 37	Construction services; installation and repair services; mining, oil, and gas drilling; The terms 'installation and repair services' need to be specified more precisely, as it is not sufficiently clear and unambiguous.
Class 38	Telecommunications services
Class 39	Transport services; packaging and storage of goods; organization of travel
Class 40	Material processing; recycling of waste and trash; air purification and water treatment; printing services; preservation of food and beverages; The term 'material processing' needs to be specified more precisely, as it is not sufficiently clear and unambiguous.
Class 41	Education; training; entertainment; sporting and cultural activities
Class 42	Scientific and technological services as well as research and related design services; industrial analysis, industrial research, and services of an industrial designer; quality control and authentication services; design and development of computer hardware and software
Class 43	Services for providing food and accommodation to guests
Class 44	Medical services; veterinary services; health and beauty care for humans and animals; services in the fields of agriculture, aquaculture, horticulture, and forestry
Class 45	Legal services; security services for the physical protection of tangible property or individuals; personal and social services provided by third parties to meet individual needs; The terms 'personal and social services provided by third parties to meet individual needs' need to be specified more precisely, as it is not sufficiently clear and unambiguous.

but must be done manually, which means that rapid trademark registration is no longer possible.

Example

The software company Best Software GmbH would like to protect its trademark "Pasquale" for its software products and software services. For Best Software GmbH, the goods software, software packages, data processing software, downloadable software in Nice class 9, and the services software consulting, software creation, software engineering, software development, and software creation services in Nice class 42 are recommended.◄

13.3 Practical Examples

Examples from typical fields of application are presented to illustrate the fundamental approach to determining goods and services.

In practice, it has proven effective to select three classes for a trademark application and to specify five to ten terms for each class. This results in a cost advantage, as in many countries three classes are included in the application fee, for example in Germany. Registering a trademark for a large number of classes does not make sense, as the trademark must be used for all classes. Otherwise, the trademark becomes subject to cancellation for those classes or for the relevant goods and services. The additional class fees would then be wasted. Furthermore, a high number of classes increases the risk of confusion with an earlier trademark, which means that by choosing a large number of classes, you incur a high risk of legal or patent office disputes.

> ▶ **Tip** As a rule, it is advisable to register a trademark for three classes. For each class, for example, five to ten specific goods or services can be specified. In most cases, it is recommended to indicate two goods classes and one service class, or one goods class and two service classes.

The Nice Classification comprises 34 classes for goods (Nice Classes 1 to 34) and 11 classes for services (Nice Classes 35 to 45). For each Nice Class, the specific goods and services for which the trademark is to be registered must be named. Only a single term can be selected for each class. However, it is advisable to specify several terms for each class. This approach can prove particularly useful later if an opposition is filed against the trademark. In such a case, it may be possible, by waiving individual goods and services, to reach a coexistence agreement with the owner of the earlier trademark without losing your own trademark entirely.

The possibility of negotiating a coexistence agreement with the owner of an earlier trademark is another reason to specify not just one, but for example three Nice Classes for a trademark application. However, it should be borne in mind that significantly more than three classes increase the likelihood of an opposition or cancellation proceeding against the trademark.

> ▶ **Tip** When specifying a Nice Class, five to ten goods or services from that class should be named.

The listing of Nice Classes is done according to their significance. When registering the trademark, the Nice Classes must be indicated in ascending order.

13.3.1 Confectionery

The Süß-und-Lecker confectionery wants to advertise its marzipan cream cake with a trademark. The trademark will be called "Lerus" and can be registered for the following classes:

- Nice Class 30 for bakery and confectionery goods, chocolate and desserts, in particular pastries, cakes, tortes and cookies, sweets, chocolate bars and chewing gum
- Nice Class 35 for retail and wholesale services relating to foodstuffs, shipping of retail goods
- Nice Class 21 for crockery, cookware and containers, in particular kitchen strainers, cake molds, baking trays and cake plates

13.3.2 Restaurant

An Italian restaurant wishes to register "Bella Mama" as a trademark. The following classes can be recommended:

- Class 43 for services for the provision of food and accommodation for guests
- Class 30 for prepared or preserved foods for consumption, in particular pizzas, pizza sauces, pizza dough, pizza seasonings, ready-made pizza meals, spaghetti and spaghetti sauce
- Class 39 for transport services, in particular delivery of food by restaurants

13.3.3 Ice Cream Parlor

The ice cream parlor "Lombisto" wants to register its business name as a trademark. The trademark "Lombisto" can be recommended for the following classes:

- Nice Class 30 for prepared or preserved foods for consumption, in particular ice cream, iced tea, iced coffee, ice confectionery, cakes, cake dough,
- Nice Class 43 for accommodation and catering for guests, catering in restaurants and services for the accommodation of guests, in particular catering
- Nice Class 39 for transport services; packaging and storage of goods, in particular delivery of food by restaurants

13.3.4 Beverage Manufacturer

BestNade GmbH has developed an innovative beverage, which is offered to the market both in a non-alcoholic version and as an alcoholic drink. The trademark is called "Nadama". The following goods and services can be used:

- Nice Class 32 for beer and brewery products, non-alcoholic beverages, in particular flavored, carbonated drinks, juices, especially vegetable juices, isotonic drinks and tomato juice, waters, especially mineral waters
- Nice Class 33 for alcoholic beverages, except beer, spirits, wines, alcoholic mixed drinks
- Nice Class 39 for transport and delivery of goods, in particular delivery of gift baskets with food and beverages and delivery of prepared food and beverages for consumption

13.3.5 Online Wine Trade

Max Kleinmann operates a lucrative wine business via Amazon and Ebay. To protect himself against increasing competition, he wants to build a trademark "Amor" for his wine business. His trademark "Amor" can be registered for the following classes at the patent office's trademark register:

- Nice Class 35 for retail and wholesale services relating to foodstuffs, shipping of retail goods
- Nice Class 33 for alcoholic beverages, in particular wines, brandy, wine vinegar, wine punch, wine-based beverages, in particular wine spritzers, and reduced-alcohol wines
- Nice Class 21 for wine bottle coolers, wine bottle openers, wine glasses and wine jugs

13.3.6 Cosmetics Company

BelArte GmbH has developed a skincare product and wants to register a trademark "Yahuda". The following classes are recommended for them:

- Nice Class 3 for non-medical products for body and beauty care, in particular skincare products, cosmetic skincare products, anti-aging skincare products, cosmetics, skincare lotions, facial toners, eyeliners, mousse and tanning oils
- Nice Class 44 for hygiene and beauty care for humans, in particular consulting services relating to cosmetics, cosmetic treatments for face and body
- Nice Class 21 for cosmetic, hygiene and beauty care utensils

13.3.7 Travel Company

Hans & Fritz Busunternehmen GbR wants to advertise its trips in the future with "KaiLu". The trademark "KaiLu" can be registered for the following classes:

- Nice Class 39 for organizing travel, arranging travel, conducting travel, accompanying travelers, booking services for tourist trips, organizing and arranging travel
- Nice Class 43 for services for the provision of food and accommodation for guests, in particular accommodation services for travelers
- Nice Class 35 for advertising, especially for travel

13.3.8 Automotive Repair Shop

An automotive repair shop wants to advertise its services with the trademark "Alhambra". The trademark can be registered for the following Nice Classes:

- Nice Class 37 for vehicle repair, maintenance, refueling and recharging, in particular auto body repair work and car cleaning
- Nice Class 12 for vehicles and means of transport, in particular cars and buses, parts and accessories for vehicles, especially car tires and car seat covers
- Nice Class 9 for car chargers, car batteries and car radios

13.3.9 Tire Trade

Come-and-ride GmbH exclusively sells an innovative tire development with particularly good braking properties. The trademark "Crab" is to be protected for these high-quality tires. The following classes can be included in the trademark application:

- Nice Class 35 for retail and wholesale services relating to tires for vehicles, in particular cars, passenger cars and trucks, shipping of retail goods
- Nice Class 12 for vehicles and apparatus for the transport of persons or goods by land, in particular parts and accessories for vehicles, especially wheels and tires for vehicles, tire repair accessories, tire covers, tire protection chains, tire treads, inner tubes, tire inserts
- Nice Class 7 for tire mounting machines
- Nice Class 37 for vehicle repair, maintenance, refueling and recharging, in particular tire repair, tire mounting and tire changing
- Nice Class 42 for tire inspection

13.3.10 Automotive Supplier

Müller GmbH & Co. KG manufactures pre-assembled wiring harnesses for the automotive industry. In the future, the products are to be marketed under the label “Zorx”. The following classes can be specified for the trademark “Zorx”:

- Nice Class 9 for electrical and electronic components, in particular cables, wires, wiring harnesses and electrical wiring harnesses
- Nice Class 12 for cars, parts and accessories for vehicles, especially car seats, car tires and car seat covers
- Nice Class 37 for vehicle repair, maintenance, refueling and recharging

13.3.11 Special Machinery Manufacturer

Haas GmbH has developed a feeding device, which it wants to name “Hamax”. The following classes can be specified for the trademark “Hamax”:

- Nice Class 7 for machine parts and controls for the operation of machines and engines, in particular feeding devices for machines and feeding devices for shredding machines
- Nice Class 42 for engineering services, in particular engineering services in mechanical engineering
- Nice Class 9 for software for monitoring, analyzing, controlling and executing processes in the physical world, in particular software for mechanical engineering

13.3.12 Bicycle Manufacturer

GoodRide GbR wants to secure a trademark for its new full-aluminum bicycle. The trademark is to be called “myMorningStar”. The following Nice Classes are recommended:

- Nice Class 12 for bicycles, parts and accessories for vehicles, in particular bicycle horns, bicycle wheels, bicycle helmets, bicycle frames, bicycle tires, bicycle brakes and bicycle pumps
- Nice Class 11 for bicycle lamps and bicycle indicators
- Nice Class 9 for bicycle helmets
- Nice Class 39 for bicycle rental

13.3.13 Chemical Industry Company

Cerax GmbH manufactures chemical intermediates for the chemical industry. These products are to be labeled "Ceracurax" in the future. The following Nice Classes can be registered for the trademark "Ceracurax":

- Nice Class 1 for chemical substances, chemical materials and chemical preparations as well as natural elements, in particular chemical intermediates for industry and chemical intermediates for manufacturing
- Nice Class 42 for scientific and technological services, in particular services of a chemical engineer, conducting chemical analyses, laboratory research in the field of chemistry

13.3.14 Construction Company

The construction company Hans Fritz & Söhne GbR offers a turnkey prefabricated house. This prefabricated house is to be marketed under the name "MuXXXL". The following Nice Classes can be recommended to Hans Fritz & Söhne GbR for their trademark "MuXXXL":

- Nice Class 37 for construction, assembly and demolition work, namely house construction, building maintenance and repair, interior and exterior cleaning of buildings
- Nice Class 19 for building materials and building elements, not of metal, in particular building materials and building elements made of sand, stone, rock, clay, minerals, concrete, wood, wood substitutes, doors, gates, windows and window coverings
- Nice Class 17 for insulating, damping and barrier materials, in particular articles and materials for thermal insulation, articles and materials for electrical insulation and articles and materials for soundproofing

13.3.15 Roofer

Master roofer Maier wants to offer his services under the trademark "Sorom". The following classes can be recommended to him:

- Nice Class 37 for construction services, namely roofing, roof covering and roofing work, installation work and repair services, namely roof repair, roof repair work and roof patching work
- Nice Class 6 for building materials and building elements made of metal, in particular roof sheet metal, roof angle iron, metal roof panels and metal gutters

- Nice Class 19 for building materials and building elements not made of metal, in particular roofing membranes, roofing sheets, roof cement, roof slate, roof shingles, roof boarding, roofing materials, tarred roofing felt, bitumen roofing felt and asphalt roofing felt

13.3.16 Plumber

Plumber Müller wants to register the trademark "Rex" for his company and use it to advertise his services. The following classes can be recommended:

- Nice Class 37 for construction services; installation work, namely installation of pipelines, repair services, namely maintenance of sanitary technology and installation of sanitary facilities
- Nice Class 11 for sanitary installations, sanitary ceramics, sanitary fittings, sanitary water fittings, sanitary valves, in particular valves, drainpipes and taps, sanitary ware made of stoneware, stainless steel or porcelain, flush tanks as sanitary devices and siphon inserts for sanitary ceramics
- Nice Class 6 for building materials and building elements made of metal, in particular pipes, tubes and hoses and accessories therefor, including valves

13.3.17 Architect

Architect Jost Baron wants to register a trademark for his architectural office. The trademark is to be called "Yoorboon". The trademark "Yoorboon" can be registered for the following classes:

- Class 42 for architectural and urban planning services, in particular services of an architect, architectural services, architectural consulting, architectural advisory services, architectural planning services and computer-aided design services relating to architecture
- Class 19 for building materials and building elements, not of metal, in particular wood for building and chipboard for building
- Class 6 for building materials and building elements made of metal, in particular metal parts for architectural purposes

13.3.18 Textile Company

TextilBest GmbH manufactures a new textile with a dirt-repellent effect. This unique selling point is to be highlighted by the trademark "Mariba". The trademark "Mariba" can be registered for the following classes:

- Nice Class 24 for fabrics, textiles and textile substitutes, in particular furniture covers, curtains and drapes, wall hangings and household linen, especially kitchen linen and table linen, bed linen and blankets and bath linen
- Nice Class 40 for material processing and transformation, namely textile, leather and fur processing, especially impregnation of textiles, waterproofing of textiles, dyeing of textiles, finishing of textiles and laminating of textiles
- Nice Class 7 for material manufacturing and processing machines, namely textile manufacturing machines and cleaning machines for textiles

13.3.19 Online Clothing Retail

Bekleidungsmanufaktur GmbH sells its own collection of men's clothing online. The new spring collection is to be promoted under the trademark "Softex". The following classes are suitable for a trademark application:

- Nice Class 35 for retail and wholesale services relating to clothing and textiles, shipping of retail goods
- Nice Class 25 for headgear and clothing items
- Nice Class 26 for clothing accessories, in particular artificial pearls as clothing trim, silver embroidery for garments, eyelets for clothing and brooches, sewing items and decorative textile articles

13.3.20 Software Company

Best Software GmbH has developed new financial accounting software for medium-sized businesses and wants to protect the trademark "Pasqale". The following classes are recommended:

- Nice Class 9 for the goods software, in particular accounting software, software packages, downloadable software, communication and network software as well as software for social networks, software for data and file management as well as for databases, office and business application software

- Nice Class 42 for software development, software engineering, rental of computer hardware and systems, IT consulting, information and advisory services and IT security, protection and maintenance services
- Nice Class 35 for computer-based accounting, computer-based accounting services, bookkeeping, accounting and auditing

13.3.21 Web Designer

GoodLook GmbH offers its clients highly professional web presences tailored to their specific needs. Recently, a do-it-yourself kit for simple websites has been offered. The name "Alhambra" was developed for this new product. The following classes can be recommended:

- Nice Class 9 for application software, communication and network software as well as software for social networks, web applications and server software, content management software
- Nice Class 42 for creation of internet websites, design of internet websites and hosting of internet portals
- Nice Class 38 for telecommunications services, internet-based telecommunications services, provision of access to content, web pages and internet portals

13.3.22 Computer Retail

Best Software GmbH operates a small chain of retail stores and offers high-quality consulting and the sale of computers with suitable application software in its stores. In the future, Best Software GmbH wants to offer its services under the trademark "Zeus". The following goods and services can be specified for the trademark application:

- Nice Class 35 for retail and wholesale services, in particular retail services for computer software, retail services relating to computer hardware and computer software
- Nice Class 42 for software consulting, software development, software engineering and software creation services
- Nice Class 9 for computers, computer mice, computer parts, computer cables, computer servers, computer printers, computer monitors, computer hardware, computer software and computer networks, software, software packages

13.3.23 Online Toy Retail

Best Toy GmbH operates an online toy store. Best Toy GmbH has decided to introduce a trademark to differentiate itself from the competition and to avoid being confused with competing companies. It turned out that MyBestToy GmbH provides poor service and apparently Best Toy GmbH is being confused with MyBestToy GmbH. As a result, Best Toy GmbH suffered from the poor reputation of MyBestToy GmbH. The new trademark is to be called "Grip" and can be registered for the following classes at the patent office:

- Nice Class 35 for retail and wholesale services relating to toys, shipping of retail goods
- Nice Class 28 for toys, games and playthings, in particular toy sets, toy buckets, toy animals, toy boats, toy balls, toy cars
- Nice Class 20 for furniture and furnishings, in particular toy chests

13.4 Specifying with "Namely"

A specification of goods or services can be made, for example, to distinguish oneself from an earlier trademark.

Example

A trademark "Alhambra" is registered in Nice Class 9 for application software for medical or surgical applications. A later trademark "Alhambris" for application software leads to a risk of confusion. A specification of the application software by indicating the particular application can eliminate the risk of confusion or at least lead to the owner of the earlier trademark accepting coexistence of the two trademarks. For example, the later trademark "Alhambris" can be limited to "for application software, namely for painting robots" and thus a conflict with the earlier right can be excluded.◀

By adding "namely", the relevant good or service is specified. However, this results in a restriction of the scope of protection of the trademark to this particular good or service.

13.5 Including Examples with "In Particular"

By adding "in particular" after a good or service, examples can be given for these goods and services. An "in particular" addition is therefore not restrictive, as is the case with a "namely" addition. An "in particular" addition can, in the course of a waiver or restriction, be converted into a "namely" addition. The reverse is not possible.

13.6 Disclaimer

Individual goods and services to be named can be excluded from a specification of goods or services by "except" or "excluding".

Example

The trademark "Alhambra" is registered in Nice Class 9 for application software for medical or surgical applications. A later trademark "Alhambris" for application software leads to a risk of confusion. A specification of the application software with a disclaimer could eliminate the risk of confusion. For example, the trademark "Alhambris" could be limited to "for application software, except for medical or surgical applications".◄

13.7 Coexistence Thanks to Dissimilar Goods and Services

A trademark is not just the brand representation or the sign itself, but a trademark consists of the combination of the sign with specific goods and services. A trademark must always be considered in connection with the registered goods and services. The trademark as a combination of a sign and the associated goods and services can be illustrated using the example of the "Duplo" trademark. The "Duplo" trademark was registered for two applicants.

Registration of the word/picture mark "Duplo"[5] for LEGO Juris A/S, 7190, Billund, Denmark, for the following goods and services (see Fig. 13.1):

- **Class 9** for scientific, nautical, surveying, electrical, photographic, cinematographic, optical, weighing, measuring, signaling, checking, lifesaving, and teaching apparatus and instruments, all as included in Class 9; apparatus for recording, transmission, and reproduction of sound and images; magnetic recording disks; recorded computer programs and software; vending machines and mechanisms for coin-operated apparatus; cash registers, calculating machines, data processing equipment, and computers; fire extinguishers; spectacles; spectacle cases; magnets
- **Class 16** for paper, cardboard, and goods made from these materials (as far as included in Class 16); printed matter; bookbinding material; photographs [prints]; stationery and office requisites; adhesives for stationery or household purposes; artists' materials; paintbrushes; teaching and instructional materials [except apparatus]; plastic packaging materials (as far as included in Class 16); playing cards

[5] EUIPO, "https://www.tmdn.org/tmview/welcome#/tmview/detail/EM500000000500306", accessed on June 21, 2021.

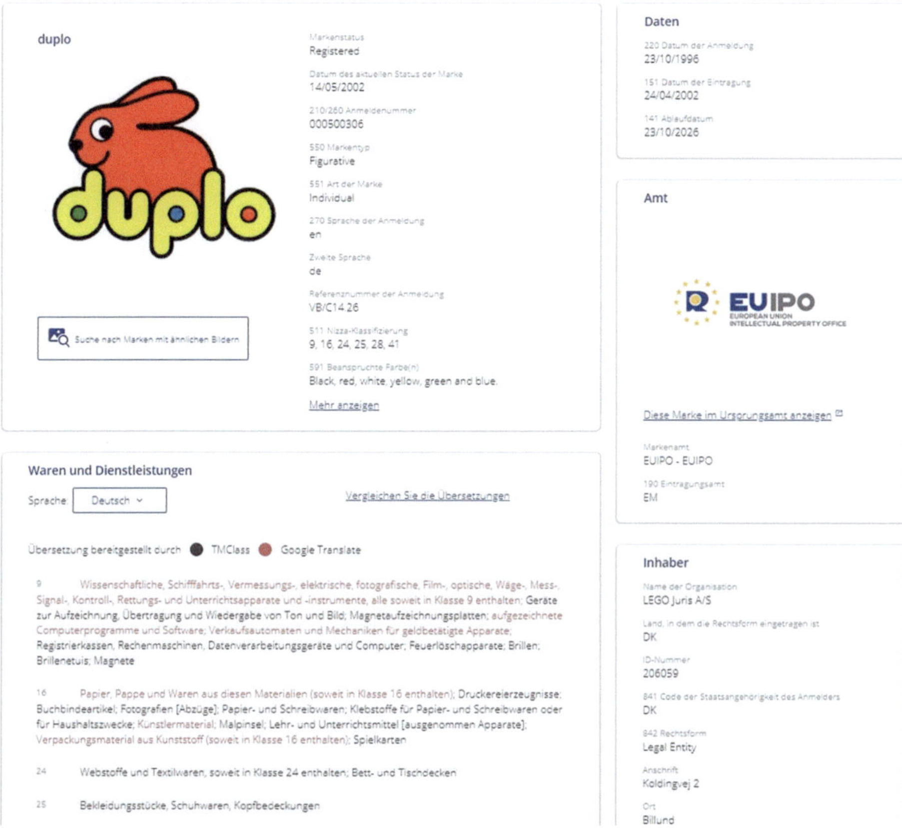

Fig. 13.1 "Duplo" trademark for Lego (EUIPO)

- **Class 24** for woven fabrics and textile goods, as far as included in Class 24; bed and table covers
- **Class 25** for clothing, footwear, headgear
- **Class 28** for games, toys; gymnastic and sporting articles (as far as included in Class 28); Christmas tree decorations
- **Class 41** for education and instruction; training; entertainment services; cultural and sporting activities; video tape and film production; amusement park services; publication of books and texts (except advertising texts).

In addition, the word/picture mark "duplo"[6] of Ferrero Deutschland GmbH (see Fig. 13.2) was registered for the following goods:

[6] EUIPO, "https://www.tmdn.org/tmview/welcome#/tmview/detail/DE500000000959526", accessed on 4.10.2021.

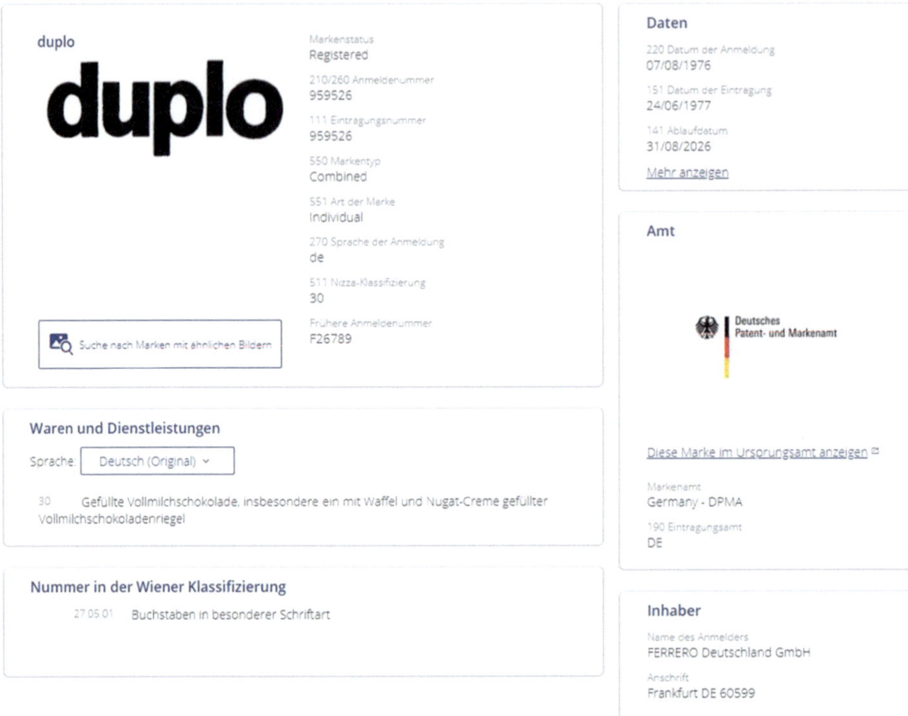

Fig. 13.2 "Duplo" trademark for Ferrero (EUIPO)

- **Class 30** for filled milk chocolate, in particular a milk chocolate bar filled with wafer and nougat cream

The two identical text elements of the word/picture marks "Duplo" are used for goods that are not similar. There is no likelihood of confusion, and the two trademarks can coexist without causing market confusion among the relevant public.

Another example of two trademarks owned by different proprietors with identical text elements, which illustrates the importance of goods and services, is the "Trumpf" trademarks.

Figure 13.3 shows the word/picture mark of Trumpf GmbH + Co. KG[7] in Ditzingen. The company is a leading manufacturer of machine tools. Accordingly, the "Trumpf" trademark is registered for the following Nice classes:

[7] EUIPO, "https://www.tmdn.org/tmview/welcome#/tmview/detail/WO500000000540412", accessed on 4.10.2021.

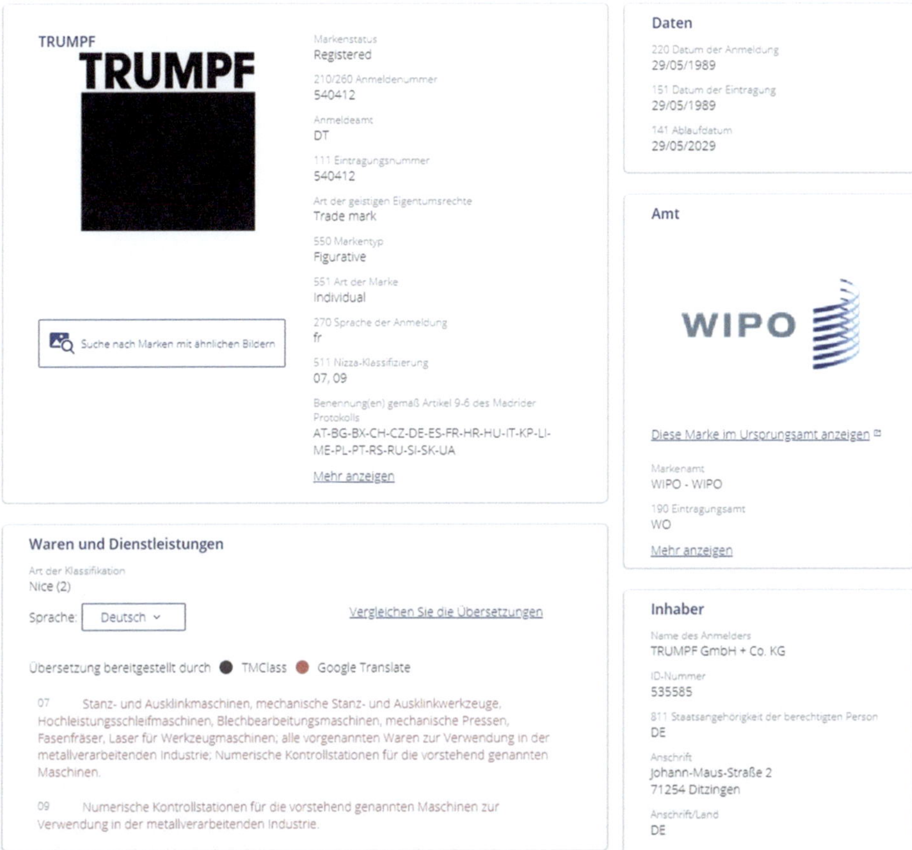

Fig. 13.3 "Trumpf" trademark for Trumpf GmbH & Co. KG (EUIPO)

- **Class 7** for punching and notching machines, mechanical punching and notching tools, high-performance grinding machines, sheet metal processing machines, mechanical presses, beveling cutters, lasers for machine tools; all the aforementioned goods for use in the metalworking industry; numerical control stations for the aforementioned machines
- **Class 9** for numerical control stations for the aforementioned machines for use in the metalworking industry

In addition, Ludwig Schokolade GmbH & Co. KG holds a word/picture mark "Trumpf"[8] (see Fig. 13.4) for the following Nice classes:

[8] EUIPO, "https://www.tmdn.org/tmview/welcome#/tmview/detail/EM500000012562443", accessed on June 21, 2021.

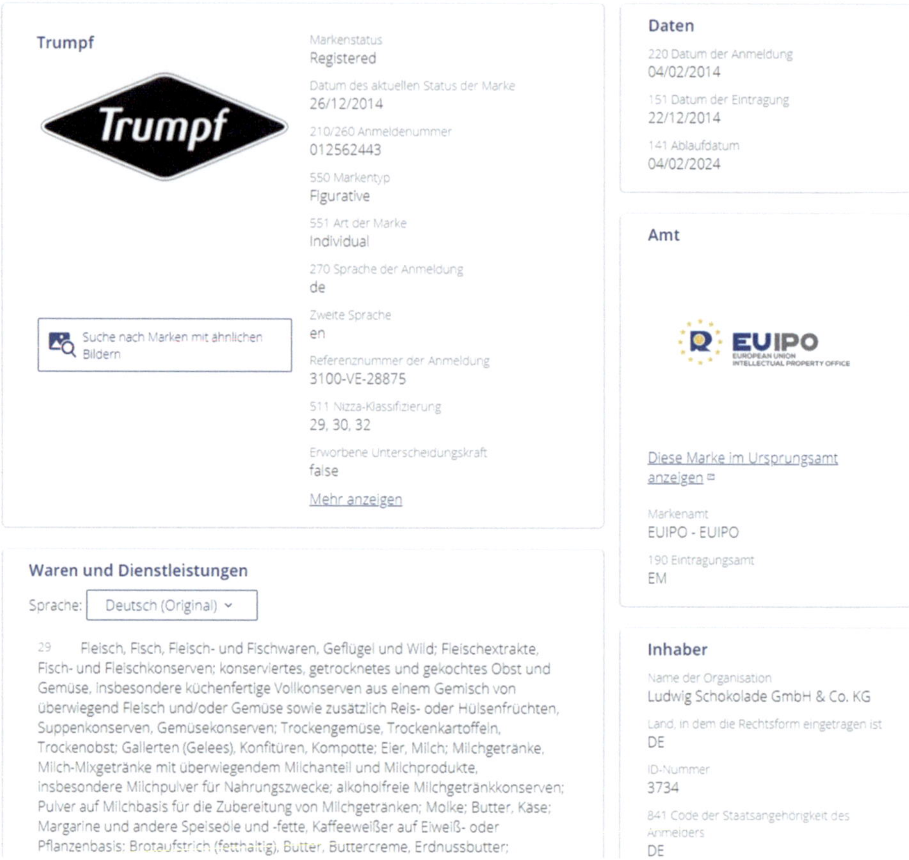

Fig. 13.4 "Trumpf" trademark of Ludwig Schokolade GmbH & Co. KG (EUIPO)

- **Class 29** for meat, fish, meat and fish products, poultry and game; meat extracts, fish and meat preserves; preserved, dried, and cooked fruits and vegetables, in particular ready-to-eat canned foods consisting mainly of meat and/or vegetables as well as rice or pulses, soup preserves, vegetable preserves; dried vegetables, dried potatoes, dried fruit; jellies, jams, compotes; eggs, milk; milk beverages, milk-based mixed drinks with a predominant milk content and dairy products, in particular milk powder for nutritional purposes; non-alcoholic milk beverage preserves; milk-based powders for the preparation of milk beverages; whey; butter, cheese; margarine and other edible oils and fats, coffee whitener based on protein or plants; spreads (fat-based), butter, buttercream, peanut butter; peanuts (processed); fruits (preserved, cooked, deep-frozen, candied, or in alcohol); fruit jellies; fruit pulp; fruit salad; jellies for food purposes; jellies for food purposes; vegetables (preserved, cooked, deep-frozen); yogurt; cocoa butter; kefir; coconut butter; coconut fat; coconuts (dried); coconut oil; compotes;

jams; almonds and nuts (processed); marmalades; fruit salad; plant juices for cooking; raisins; whipped cream; edible gelatine; edible tallow; the aforementioned goods in Class 29 also as dietary foods for non-medical purposes and, where possible, also in instant form; fruit slices

- **Class 30** for coffee; coffee substitutes; tea, fruit tea, and herbal tea for non-medical purposes; cocoa and cocoa powder, caffeinated, cocoa-containing and/or chocolate-containing beverage powders; coffee beverages, tea beverages, cocoa beverages; drinking chocolate; cappuccino beverages and beverage powders; sweet spreads, namely nut-nougat creams and chocolate creams; nut pastes; chocolate and chocolate products, in particular chocolate bars, also in the form of portioned individual pieces; pralines, also with liquid filling, especially made from wines and spirits; muesli and chocolate bars, in particular filled bars, also with caramel and/or nuts and/or nut pieces; fine baked goods (sweet and savory) and confectionery; long-life baked goods, including those with coatings of fat glazes or chocolate and nut or almond pieces; confectionery, in particular chewy candies, including fruit chewy candies; molded chocolate and confectionery, in particular figures and figure assortments; sugar; dragees (confectionery) and chewing gum for non-medical purposes; filled chocolate; hollow chocolate figures with toys and/or small toys inside; ice cream; rice, groats for food purposes; flours and cereal preparations, pasta; bread; honey; baking powder; pudding preserves; non-alcoholic cocoa beverage preserves; all the aforementioned goods in Class 30 also for dietary, non-medical purposes and, where possible, also in instant form
- **Class 32** for beers; mineral waters, carbonated waters, and other non-alcoholic beverages; waters with added caffeine, tea, or cocoa; fruit drinks, fruit juices, and vegetable juices, fruit juice beverages; isotonic beverages; syrups and other preparations for making beverages, in particular effervescent tablets and beverage powders for the preparation of non-alcoholic or isotonic beverages (as far as included in Class 32); all the aforementioned goods in Class 32 also for dietary non-medical purposes and, where possible, also in instant form

Note

A trademark is to be understood as a combination of a sign, the brand representation, and the associated goods and services.◄

13.8 Accelerating Trademark Registration

With a few clever moves, the registration of a trademark can be accelerated. In any case, the application fee should be paid directly when submitting the application documents. The patent office will only begin processing the application documents after receipt of payment. Of course, formal errors that result in official notifications for correction should be avoided. Particular attention should be paid to the correct indication of the applicant. In addition, the specification of goods and services should be made in accordance with the requirements of the Nice Classification. The specifications of goods and services described in TMclass[9] are recommended. The European patent offices and the EUIPO have agreed on these specifications for goods and services.

13.8.1 Lead Class

One of the classes must be designated as the so-called lead class. The applicant is encouraged to choose as the lead class the one that represents the future economic focus of the use of the trademark. Based on the choice of the lead class, the patent office will generally determine which trademark office is responsible for processing the trademark application.

By choosing the lead class, the applicant can therefore determine the responsible trademark office. It should be noted that there are trademark offices with a high workload and others with a low workload. By choosing the lead class, the applicant can direct their application to a trademark office with a low workload and thus enable a quick registration of the trademark.

In the past, rapid processing of trademark applications was particularly achieved for lead class 22. Class 22 mainly includes raw textile fibers and their substitutes, goods made of fabric and fibers, and upholstery and filling materials. A quick processing was also guaranteed in the past with lead class 34. Class 34 mainly includes smokers' articles.

Registration can be accelerated by adding class 22 to a trademark application. In this case, a fee of € 100 is due for the additional class. Class 22 should be chosen as the lead class. After the trademark has been registered, class 22 can be dropped.

13.8.2 Request for Acceleration

The applicant can file a request for accelerated examination. The application will then be processed with priority, and the patent office guarantees, with the applicant's cooperation, to decide on the application within six months. This means that the trademark will be

---

[9] EUIPO, "http://euipo.europa.eu/ec2/?lang=de", accessed on May 31, 2021.

registered within six months, provided there are no formal deficiencies or absolute grounds for refusal. A fee of € 200 is due for a request for acceleration.[10]

It can be assumed that, in any case, an examination of the trademark application will take place within this period. For this reason, a request for acceleration is less advisable. As a rule, a normal processing time of six to twelve weeks can be expected.

13.9 Trademark Symbols ®, © and TM

The trademark symbols ®, ©, and TM originate from Anglo-American law. The use of these symbols is permitted in Germany, but the symbols must not be used in a misleading manner. Misleading use can be subject to a warning. If the trademark symbols are used correctly, a positive advertising effect may result. In addition, the trademark symbols can deter potential infringers, possibly avoiding opposition and cancellation proceedings.

The trademark symbols are not to be included in the brand representation. The use of the trademark symbols ® and TM would then, at least temporarily, be misleading, as the symbols indicate temporally successive states of a trademark. The designation would therefore be abusive at the beginning or at a later point in time. The third symbol © has nothing to do with trademark law and should therefore not be used in connection with a trademark at all.

13.9.1 R in a Circle ®: Registered

The symbol ® means that the marked trademark is registered in the trademark register. The ® symbol must therefore not be used for a trademark that has not yet been registered, but only applied for. Under German trademark law, a registered trademark may be marked with the ® symbol to signal to the relevant public that trademark protection exists.

It should be ensured that a brand representation is only marked with the ® symbol in the manner in which it is registered in the register, i.e., completely and not in a distorted form. Therefore, the entire trademark should be marked with the ® symbol and not just individual components, for example, only the word element of a word/picture mark.

An impermissible use of the ® symbol also occurs if the marked trademark is used in connection with goods and services for which the trademark is not registered.

13.9.2 TM: Trademark

The addition "TM" in the Anglo-American legal system designates trademarks that have been applied for but not yet registered in the trademark register. In Germany, the legal

[10] DPMA, "https://www.dpma.de/marken/anmeldung/index.html", accessed on June 11, 2021.

situation is inconsistent. The Munich Regional Court considers the use of the TM symbol to be misleading if the symbol is used for an applied-for trademark that has not yet been registered.[11] The Munich Regional Court thus equates the meaning of the TM symbol with that of the ® symbol. In contrast, the Essen Regional Court does not consider the use of the TM symbol for a merely applied-for trademark to be misleading.[12] The use of the TM symbol therefore carries a certain risk potential.

13.9.3 C in a Circle ©: Copyright

The symbol © denotes a copyright and not a trademark right. Copyright arises solely through the creation of a work and does not need to be entered in a register, as is the case with a trademark, in order to take effect.

Copyright arises, for example, through the creation of a logo by a designer. The designer is the creator of the logo and can mark the logo with a © symbol. A licensee of the logo does not become the creator by taking a license. Therefore, the licensee is not permitted to use the © symbol, even with a license. Otherwise, the licensee would be presenting themselves as the author, which they are not.

13.10 Registering a German Trademark

The different options for submitting an application for a German trademark are presented.

13.10.1 Application by Mail or Fax

The application form W 7005/1.19 for a German trademark can be downloaded at the link "https://www.dpma.de/docs/formulare/marken/w7005.pdf".[13]

On the first page of the form, in box (1), the name and address to which the patent office's correspondence should be sent must be entered (see Fig. 13.5). If the applicant does not have a patent attorney as a representative, they enter their own details in box (1) and do not need to fill in box (3).

If the application is sent by fax, it can also be sent by mail. In this case, the upper box on the top right must be checked and the date of the fax transmission entered. A word mark can be submitted to the patent office exclusively by fax. In this case, the lower box must be checked. These details help the patent office to match the fax transmission and any subsequent postal submission of the same application.

[11] Munich Regional Court I, 23.7.2003, 1 HK O 1755/03.

[12] Essen Regional Court, 4.6.2003, 44 O 18/03.

[13] DPMA, "https://www.dpma.de/docs/formulare/marken/w7005.pdf", accessed on June 11, 2021.

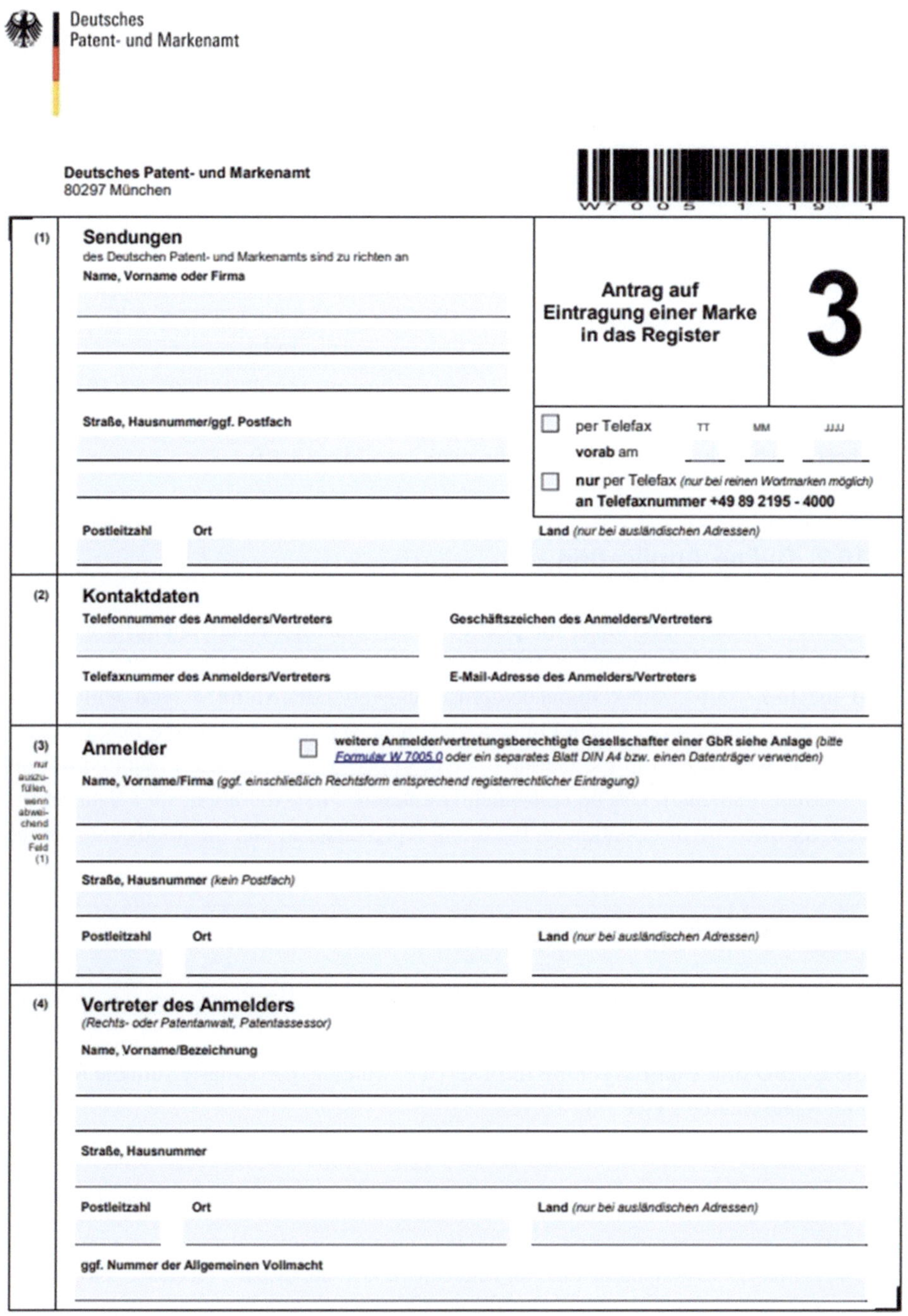

Deutsches
Patent- und Markenamt

Deutsches Patent- und Markenamt
80297 München

W7005 1.19 1

(1) **Sendungen**
des Deutschen Patent- und Markenamts sind zu richten an
Name, Vorname oder Firma

Straße, Hausnummer/ggf. Postfach

Postleitzahl **Ort** **Land** *(nur bei ausländischen Adressen)*

Antrag auf Eintragung einer Marke in das Register

3

☐ per Telefax TT MM JJJJ
vorab am

☐ **nur** per Telefax *(nur bei reinen Wortmarken möglich)*
an Telefaxnummer +49 89 2195 - 4000

(2) **Kontaktdaten**

Telefonnummer des Anmelders/Vertreters **Geschäftszeichen des Anmelders/Vertreters**

Telefaxnummer des Anmelders/Vertreters **E-Mail-Adresse des Anmelders/Vertreters**

(3) nur auszufüllen, wenn abweichend von Feld (1)

Anmelder ☐ **weitere Anmelder/vertretungsberechtigte Gesellschafter einer GbR siehe Anlage** *(bitte Formular W 7005.0 oder ein separates Blatt DIN A4 bzw. einen Datenträger verwenden)*

Name, Vorname/Firma *(ggf. einschließlich Rechtsform entsprechend registerrechtlicher Eintragung)*

Straße, Hausnummer *(kein Postfach)*

Postleitzahl **Ort** **Land** *(nur bei ausländischen Adressen)*

(4) **Vertreter des Anmelders**
(Rechts- oder Patentanwalt, Patentassessor)

Name, Vorname/Bezeichnung

Straße, Hausnummer

Postleitzahl **Ort** **Land** *(nur bei ausländischen Adressen)*

ggf. Nummer der Allgemeinen Vollmacht

Fig. 13.5 Application form DE trademark page 1 (DPMA)

On the second page of the application form, a word mark can be entered in box (5) at the top. Below that, it can be indicated that an additional sheet or form W 7005.1/1.19 is attached, on which the representation of a figurative mark or a word/figurative mark is shown. The type of mark is specified in box (6) (see Fig. 13.6).

On the third page, in box (9), the Nice classes and the desired terms from the Nice classes must be entered. If a priority is claimed, this can be noted in box (11) (see Fig. 13.7).

On the fourth page of the application form, a request for accelerated examination can be made in box (12). An accelerated examination incurs an official fee of € 200. In box (14), the payment method can be selected (see Fig. 13.8).

On the fifth page, the application must be signed without fail. Otherwise, no valid application will be submitted to the patent office (see Fig. 13.9).

On the supplementary form "Annex to the application form" W 7005.1/1.19[14] a figurative mark or a word/figurative mark can be depicted. Alternatively, an additional DIN A4 sheet can be attached to represent the mark (see Fig. 13.10).

13.10.2 Online Application

The German Patent Office offers the possibility of online trademark application. The use of the DPMAdirektWeb service is possible without a signature card and card reader.

At the link "https://direkt.dpma.de/dpmadirektweb/index.xhtml"[15] you will find the first screen of the German Patent Office's trademark application tool (see Fig. 13.11).

With this web tool from the German Patent Office, the information required to apply for a German trademark can be conveniently entered. First, the applicant's data must be entered, followed by information on the brand representation and the goods and services.

13.11 Registering a European Union Trademark

At the EUIPO, application by fax or post is no longer possible. Therefore, only online application for a European Union trademark is permitted.

The EUIPO offers an EasyFiling form service, which is particularly suitable for SMEs without representatives (see Fig. 13.12).[16]

[14] DPMA, "https://www.dpma.de/docs/formulare/marken/w7005_1.pdf", accessed on June 11, 2021.

[15] DPMA, "https://direkt.dpma.de/dpmadirektweb/index.xhtml", accessed on June 21, 2021.

[16] EUIPO, "https://euipo.europa.eu/ohimportal/de/apply-now", accessed on June 16, 2021.

W7005 1.19 2

(5) **Markendarstellung** *(pro Anmeldung nur eine Marke)*

☐ ____________________
(Wortmarke)

oder

☐ **siehe Anlage** *(Anlage zwingend erforderlich für alle Markenformen, ausgenommen Wortmarken; bitte Formular W 7005.1 oder ein separates Blatt DIN A4 bzw. einen Datenträger verwenden)*

! Ein ® sollte der Markendarstellung nicht schon bei der Anmeldung hinzugefügt werden, da unter Umständen eine Zurückweisung wegen Täuschungsgefahr gemäß § 8 Abs. 2 Nr. 4 Markengesetz in Betracht kommen kann.

(6) **Markenform**

☐ **Wortmarke** *(Wörter, Buchstaben, Zahlen, sonstige Schriftzeichen ohne grafische und/oder farbige Ausgestaltung)*

☐ **Wort-/Bildmarke** *(Kombination aus Wort und/oder Zahl und Bild, grafisch und/oder farbig gestaltete Wörter)*

☐ **Bildmarke** *(zweidimensionale Bilder ohne Wort- und/oder Zahlelemente)*

☐ **Dreidimensionale Marke** *(dreidimensionale Gestaltungen)*

☐ **Farbmarke** *(z.B. abstrakte Farbe oder Kombination aus mehreren Farben; als Markendarstellung Feld (5) ist ein Farbmuster einzureichen)*
Bezeichnung der Farbe/n nach einem international anerkannten Farbklassifikationssystem (z.B. RAL, Pantone, HKS)
____

☐ Beschreibung der Anordnung der Farben zueinander (räumliche Anordnung und Größenverhältnis) ist als Anlage beigefügt
(nur erforderlich bei einer Kombination der Marke aus mehreren Farben)

☐ **Klangmarke** *(akustisch hörbare Töne/Melodien; als Markendarstellung (Feld (5)) ist eine Notenschrift oder eine Wiedergabe auf einem Datenträger einzureichen)*

☐ **Andere Markenform** *(Positionsmarke, Kennfadenmarke, Mustermarke, Bewegungsmarke, Multimediamarke, Hologrammmarke, sonstige Marke)*
nämlich ____________________
(bitte nur eine der genannten Markenformen angeben)

☐ **Markenbeschreibung ist als Anlage beigefügt**
(nur erforderlich, wenn die Markendarstellung den Schutzgegenstand – in objektiver Weise – nicht hinreichend bestimmt; darf maximal 150 Wörter in einem fortlaufenden Text und keine grafischen oder sonstigen Gestaltungselemente enthalten; bitte Formular W 7005.2 oder ein separates Blatt DIN A4 bzw. einen Datenträger verwenden)

(7) **Farbangaben zur Markendarstellung** *(nicht auszufüllen bei Wortmarken, Klangmarken und Farbmarken)*

☐ Die Markendarstellung enthält **farbige** Elemente und zwar in folgenden Farben
(bitte allgemeine Farbnamen angeben, z.B. gelb, weiß, rot, grün, schwarz, blau)
____

(8) **Nichtlateinische Schriftzeichen**
(zwingend auszufüllen, wenn die Marke nichtlateinische Schriftzeichen beinhaltet)

☐ Die Markendarstellung enthält **nichtlateinische** Schriftzeichen

Transliteration (buchstabengetreue Wiedergabe) ____________
Transkription (phonetische Wiedergabe in lateinischen Schriftzeichen) ____________
Übersetzung (in die deutsche Sprache) ____________

Beispiel für „Буква"	
Transliteration	Bukva
Transkription	Bukwa
Übersetzung	Buchstabe

☐ Transliteration, Transkription und Übersetzung sind als Anlage beigefügt
(bitte Formular W 7005.3 oder ein separates Blatt DIN A4 bzw. einen Datenträger verwenden)

Fig. 13.6 Application form DE trademark page 2 (DPMA)

W7005 1.19 3

(9) **Verzeichnis der Waren und/oder Dienstleistungen**

(zwingend auszufüllen)

☐ Verzeichnis der Waren und/oder Dienstleistungen ist als Anlage beigefügt (*bitte Formular W 7005.4 oder ein separates Blatt DIN A4 bzw. einen Datenträger verwenden*)

Bitte gruppieren Sie Ihr Verzeichnis nach Klassen und trennen Sie die einzelnen Waren und/oder Dienstleistungen innerhalb der angegebenen Klassen durch Semikola voneinander. Nutzen Sie für die Erstellung, soweit möglich, die harmonisierten und zulässigen Begriffe der einheitlichen Klassifikationsdatenbank (eKDB).

Leitklassenvorschlag des Anmelders ______

Klassen	**Waren und/oder Dienstleistungen** *(zwingend zu benennen; nur Angabe der Klassen nicht ausreichend)*

(10) **Serienanmeldung**

☐ Die Anmeldung ist Bestandteil **einer Serie** von Markenanmeldungen (*Vorblatt W 7002 bitte zwingend ausfüllen und beifügen*)

☐ Die Serie enthält identische Waren-/Dienstleistungsverzeichnisse

Diese Anmeldung ist Nummer ______ von ______ Anmeldungen

(11) **Priorität**

☐ **Ausländische Priorität**
Kopie/Abschrift der ausländischen Voranmeldung

☐ ist beigefügt

☐ wird nachgereicht

Datum	**Staat**	**Aktenzeichen**

☐ **Ausstellungspriorität** ☐ **Ausstellungsbescheinigung** (*Formular W 7708 bitte ausfüllen und beifügen*)

Bezeichnung der Ausstellung

______

Fig. 13.7 Application form DE trademark page 3 (DPMA)

W7005 1.19 4

(12) **Sonstige Anträge**

☐ Antrag auf **beschleunigte Prüfung** nach § 38 Markengesetz *(gebührenpflichtig)*

☐ Antrag auf Eintragung als **Kollektivmarke** nach §§ 97 ff. Markengesetz *(nicht für Privatpersonen möglich; Markensatzung zwingend erforderlich – vgl. Feld (15), Anlagen)*

☐ Antrag auf Eintragung als **Gewährleistungsmarke** nach §§ 106a ff. Markengesetz *(nicht für Hersteller/Lieferanten möglich; Markensatzung zwingend erforderlich - vgl. Feld (15), Anlagen)*

☐ Antrag auf **internationale Registrierung** dieser Markenanmeldung liegt bei *(Begleitschreiben M 8005 und Formblatt der WIPO MM2 – vgl. Feld (15), Anlagen)*

(13) **Sonstige Erklärungen**

Der Anmelder ist bereit zur

☐ **Lizenzierung der Marke** (§ 42c MarkenV)

☐ **Veräußerung der Marke** (§ 42c MarkenV)

(14) **Gebührenzahlung** in Höhe von ________ €

Zahlung per Banküberweisung

☐ **Überweisung**
(dreimonatige Zahlungsfrist beachten)

Zahlungsempfänger:
Bundeskasse Halle/DPMA
IBAN: DE84 7000 0000 0070 0010 54
BIC (SWIFT-Code): MARKDEF1700

Anschrift der Bank:
Bundesbankfiliale München
Leopoldstr. 234, 80807 München

Zahlung mittels SEPA-Basis-Lastschrift

☐ Ein gültiges **SEPA-Basis-Lastschriftmandat** (Formular A 9530)

☐ liegt dem DPMA bereits vor *(Mandat für mehrmalige Zahlungen)*

☐ ist beigefügt

☐ **Angaben zum Verwendungszweck** (Formular A 9532) des Mandats mit Mandatsreferenznummer sind beigefügt

! **Wird die Anmeldegebühr nicht innerhalb von 3 Monaten nach dem Tag des Eingangs der Anmeldung gezahlt, so gilt die Anmeldung als zurückgenommen. Bitte beachten Sie, dass die Prüfung der Schutzfähigkeit der Marke erst nach Zahlungseingang beginnt.**

(15) **Anlagen** ☐ in Papierform beigefügt ☐ auf beiliegendem Datenträger (zulässige Datenträgerformate)

☐ **Angaben zu weiteren Anmeldern** (Formular W 7005.0, *separates Blatt DIN A4-Format oder pdf-Datei)* – Feld (3)

☐ **Markendarstellung** (Formular W 7005.1, *separates Blatt DIN A4-Format oder Datei)* – Feld (5)

☐ **Markenbeschreibung** (Formular W 7005.2, *separates Blatt DIN A4-Format oder pdf-Datei)* – Feld (6)

☐ **Transliteration, Transkription und Übersetzung** (Formular W 7005.3, *separates Blatt DIN A4-Format oder pdf-Datei)* – Feld (8)

☐ **Verzeichnis der Waren und/oder Dienstleistungen** (Formular W 7005.4, *separates Blatt DIN A4-Format oder pdf-Datei)* – Feld (9)

☐ **Prioritätsunterlagen** (Formular W 7708, *separates Blatt DIN A4-Format oder pdf-Datei)* – Feld (11)

☐ **Markensatzung** *(ungebunden im DIN A4-Format oder pdf-Datei)* – Feld (12)

☐ **Antrag auf internationale Registrierung** *(Begleitschreiben M 8005 und Formblatt der WIPO MM2)* – Feld (12)

☐ ________________________________

Fig. 13.8 Application form DE trademark page 4 (DPMA)

W7005 1.19 5

(16) **Unterschrift**

Der Unterschrift ist der Name in Druckbuchstaben oder Maschinenschrift hinzuzufügen; bei Firmen die Bezeichnung entsprechend registerrechtlicher Eintragung mit Angabe der Stellung/Funktion des Unterzeichnenden.

Bitte beachten Sie hinsichtlich der Verarbeitung Ihrer personenbezogenen Daten unser Merkblatt A 9106 „Datenschutz bei Schutzrechtsanmeldungen". Dieses finden Sie unter www.dpma.de: Service – Formulare – Sonstige Formulare – Hinweise zum Datenschutz.

! Die Daten Ihrer Anmeldung werden in dem elektronischen Schutzrechtsauskunftssystem DPMAregister veröffentlicht (§ 33 Abs. 3 MarkenG).

Datum

Unterschrift/en *(bei Anmeldergemeinschaften die Unterschriften aller Anmelder),* ***ggf. Firmenstempel***

Funktion/en der/des Unterzeichner/s

Fig. 13.9 Application form DE trademark page 5 (DPMA)

Deutsches
Patent- und Markenamt

Anlage zum Anmeldeformular

W 7 0 0 5 . 1 1 . 1 9 1

Geschäftszeichen des Anmelders/Vertreters	Datum der Anmeldung
	TT MM JJJJ

Markendarstellung

Nach Eingang des Antrags beim DPMA kann die angemeldete Marke nicht mehr verändert werden.
Reichen Sie Ihre Marke daher so ein, wie sie geschützt werden soll.

! **Ein ® sollte der Markendarstellung nicht schon bei der Anmeldung hinzugefügt werden, da dies unter Umständen zu einer Zurückweisung der Anmeldung wegen Täuschungsgefahr gemäß § 8 Abs. 2 Nr. 4 Markengesetz führen kann.**

Fig. 13.10 Supplementary form W 7005.1/1.19 (DPMA)

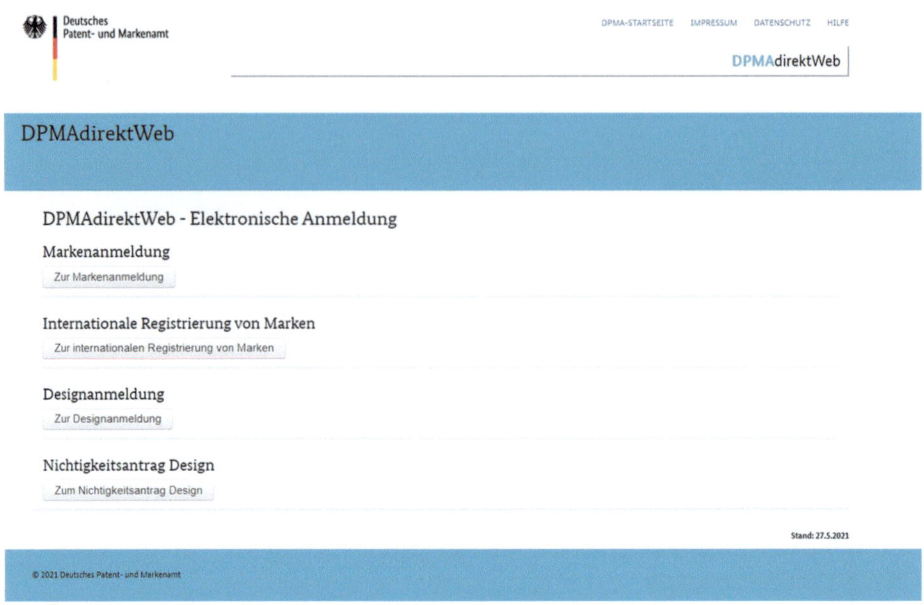

Fig. 13.11 DPMAdirektWeb (DPMA)

13.12 Filing an International Registration

An international registration can be filed directly with WIPO or with the German Patent Office. Filing with the German Patent Office is only possible if a German trademark is used as the basic mark.

13.12.1 Application by Mail or Fax

When filing an international registration with the German Patent Office, the German Patent Office's form M 8005/12.15[17] must be attached to WIPO's form MM2[18].

In form M 8005/12.15, box (1) must be completed so that the Patent Office can send correspondence to the applicant. The payment method must be entered in box (7).

If the application for international registration is a first filing, the upper left box for MM2 in box (4) must be checked. In addition, box (4) can be used to indicate whether the

[17] DPMA, Form M 8005/12.15, "https://www.dpma.de/docs/formulare/marken/m8005.pdf", retrieved on June 11, 2021.

[18] WIPO, Form MM2, "https://www.wipo.int/export/sites/www/madrid/en/forms/docs/form_mm2.pdf", retrieved on June 11, 2021.

Fig. 13.12 EasyFiling form (EUIPO)

request for international registration should be forwarded to WIPO based on the application or only after registration of the basic mark. In the case of a subsequent designation, the lower left box for MM4 in box (4) must be checked. Do not forget the signature (see Fig. 13.13).

On the first page of WIPO's MM2 application form, the applicant's name, address, and email address can be entered under Nos. 2 a, b, and c (see Fig. 13.14).

On the second page, under No. 2 f., the preferred language for correspondence with the EUIPO can be selected. English, French, and Spanish are available (see Fig. 13.15).

On the third page, the entitlement to file an IR trademark must be indicated, namely that the applicant has a residence or business establishment in a member state of the Madrid Agreement or Protocol (see Fig. 13.16).

An
Deutsches Patent- und Markenamt
80297 München

DEUTSCHES PATENT- UND MARKENAMT

(1) Name/Firma, Str./Haus-Nr., PLZ/Ort, ggf. Postf.

Sendungen des Deutschen Patent- und Markenamts sind zu richten an:

☐ **Internationale Registrierung einer Marke**
Aktenzeichen der Basismarke/-anmeldung
(bitte vollständig angeben)

IR

☐ **Nachträgliche Benennung zu einer IR-Marke**
Aktenzeichen der IR-Marke

☐ **TELEFAX** (+49 89 2195-**4000**) vorab am

(2)

Zeichen des Antragstellers/Vertreters (max. 20 Stellen)	Telefon-Nr. des Ast./Vertr.	Telefax-Nr. des Ast./Vertr.	Datum

(3) Name/Firma, Str./Haus-Nr., PLZ/Ort, ggf. Postf, wenn abweichend von Feld (1)

Antragsteller **Vertreter**

(4) **Folgendes Antragsformblatt der WIPO/OMPI ist ausgefüllt beigefügt:**

☐ MM2 Zutreffendes ankreuzen, wenn die Basismarke noch nicht im Register eingetragen ist:

☐ Das Gesuch soll aufgrund der Basisanmeldung an die WIPO/OMPI weitergeleitet werden.

☐ Das Gesuch soll erst **nach** Eintragung der Basismarke an die WIPO/OMPI weitergeleitet werden.

Werden hierzu keine Angaben gemacht, wird das Gesuch aufgrund der Basisanmeldung, also schon vor Eintragung der Basismarke an die WIPO/OMPI weitergeleitet.

☐ MM4

(5) **Dem Antrag sind folgende Formblätter beigefügt:**

☐ MM17 (Beanspruchung einer Seniorität bei Benennung der Europäischen Gemeinschaft)

☐ MM18 (Benutzungsabsichtserklärung für die USA) (Unbedingt beifügen bei Benennung USA)

(6) **Zum Antrag auf internationale Registrierung (MM2) ist**

☐ eine Abbildung der Bildmarke bzw. dreidimensionalen Marke beigefügt (in Schwarz-weiß, wenn die Basismarke schwarz-weiß eingetragen bzw. angemeldet ist; in Farbe, wenn die Basismarke farbig eingetragen bzw. angemeldet ist. Bitte Farbangaben nicht vergessen).

(Bitte zulässige Größe beachten; mindestens 1,5 cm x 1,5 cm und ***maximal 8 cm x 8 cm****)*

(7) **Gebührenzahlung** von ________ Euro

(bei MM2: Gebührennr. 334 100, **180,-- Euro**; bei MM4: Gebührennr. 334 300, **120,-- Euro**)

Die Gebühren sind innerhalb von 1 Monat nach Einreichung der Anmeldung zu zahlen.

Zahlung per Banküberweisung

☐ **Überweisung (Zahlungsbeleg ist beigefügt)**

Zahlungsempfänger:
Bundeskasse Halle/DPMA
IBAN: DE84 7000 0000 0070 0010 54
BIC (SWIFT-Code): MARKDEF1700

Anschrift der Bank:
Bundesbankfiliale München
Leopoldstr. 234, 80807 München

Zahlung mittels SEPA-Basis-Lastschrift

☐ Ein gültiges **SEPA-Basis-Lastschriftmandat** (*Vordruck A 9530*) mit der Mandatsreferenznummer *(bitte eintragen)*:

☐ liegt dem DPMA bereits vor *(Mandat für mehrmalige Zahlungen)*

☐ ist beigefügt

☐ **Angaben zum Verwendungszweck** (*Vordruck A 9532*) des Mandats mit der o. g. Mandatsreferenznummer sind beigefügt.

M 8005
12.15

Unterschrift

Fig. 13.13 IR trademark application form (DPMA)

MM2 (E) – APPLICATION FOR INTERNATIONAL REGISTRATION UNDER THE MADRID PROTOCOL

For use by the applicant:

Number of continuation sheets for several applicants:

Number of continuation sheets:

Number of MM17 forms:

☐ MM18 form (if applicable, check the box)

Applicant's reference (optional):

For use by the Office of origin:

Office's reference (optional):

1. CONTRACTING PARTY WHOSE OFFICE IS THE OFFICE OF ORIGIN

2. APPLICANT[1]

If there is more than one applicant, indicate the number of applicants and complete the "Continuation Sheet for Several Applicants".

Number of applicants:

(a) **Name:**

(b) **Address:**

(c) **E-mail address[2]:**

[1] If there is more than one applicant, indicate the details for the first applicant only and provide the details requested in the "Continuation Sheet for Several Applicants" attached to this form.

[2] You **must** indicate the e-mail address of the applicant. If a representative is appointed, the e-mail address of the applicant and of the representative must be different. If you do not indicate the e-mail address of the applicant or it is the same as the e-mail address of the representative, you will receive an irregularity notice and delay the processing of the application. The applicant must ensure that the e-mail address indicated here is correct and kept up to date.

WIPO will send all communications for this international application and the resulting international registration to the e-mail address of the applicant, **unless** an alternative e-mail address for correspondence is indicated in item 2(g)(ii) or a representative is appointed in item 4.

Where a representative is appointed, WIPO will only send communications to the e-mail address of the representative, except for a few communications where the Regulations require that WIPO send a copy to the holder (see the Note for Filing Form MM2).

MM2 (E) – May 2021

Fig. 13.14 IR trademark application form page 1 (WIPO)

MM2 (E), page 2

(d) **Telephone number[3]:**

(e) **Nationality or legal nature and State of organization[4]:**

(i) ☐ If the applicant is a **natural person**, indicate the nationality of the applicant:

Nationality of the applicant:	

(ii) ☐ If the applicant is a **legal entity**, provide **both** of the following indications:

Legal nature of the legal entity:	
State (country) and, where applicable, territorial unit within that State (canton, province, state, etc.), under the law of which the said legal entity has been organized:	

Correspondence details (optional):

(f) **Preferred language for correspondence[5]:** ☐ English ☐ French ☐ Spanish

(g) **Alternative address and e-mail address for correspondence[6]:**

(i) Postal address:

(ii) E-mail address:

[3] Indicating a phone number is not required, but it will allow WIPO to reach you if needed.

[4] Certain designated Contracting Parties may require these indications; only provide indications in either item (i) or item (ii) but **not** in both items.

[5] If you do not indicate your preferred language, WIPO will send all communications concerning this international application and the resulting international registration in the language of the international application.

[6] Use this **only** if you want WIPO to send all communications concerning this international application and the resulting international registration to an address and e-mail address different from those indicated in item 2 (b) and (c).

MM2 (E) – May 2021

Fig. 13.15 IR trademark application form page 2 (WIPO)

MM2 (E), page 3

3. ENTITLEMENT TO FILE[7]

(a) Check the appropriate box:

(i) ☐ where the Contracting Party mentioned in item 1 is a State, the applicant is a national of that State; or

(ii) ☐ where the Contracting Party mentioned in item 1 is an organization, the name of the State of which the applicant is a national: ______________________; or

(iii) ☐ the applicant is domiciled in the territory of the Contracting Party mentioned in item 1; or

(iv) ☐ the applicant has a real and effective industrial or commercial establishment in the territory of the Contracting Party mentioned in item 1.

(b) Where the address of the applicant, given in item 2(b), is not in the territory of the Contracting Party mentioned in item 1, indicate in the space provided below:

(i) if the box in paragraph (a)(iii) of the present item has been checked, the domicile of the applicant in the territory of that Contracting Party, or,

(ii) if the box in paragraph (a)(iv) of the present item has been checked, the address of the applicant's industrial or commercial establishment in the territory of that Contracting Party.

[7] If there is more than one applicant, indicate the entitlement details for the first applicant only and provide the details requested in the "Continuation Sheet for Several Applicants" attached to this form.

MM2 (E) – May 2021

Fig. 13.16 IR trademark application form page 3 (WIPO)

On the fourth page, the basic mark can be indicated under No. 5. In addition, a priority claim can be made under No. 6 (see Fig. 13.17).

On the fifth page, the trademark representation must be entered under No. 7 (see Fig. 13.18).

If a word mark or a word/device mark is being registered, it may be necessary to provide a translation under No. 9 b (see Fig. 13.19).

On the seventh page, further information on the trademark representation must be provided (see Fig. 13.20).

On the eighth page, the Nice classes and the goods and services must be entered (see Fig. 13.21).

On the ninth page, the countries for which trademark protection is to be extended must be specified. For example, extension to a European Union trademark is done by checking the EM box (see Fig. 13.22).

On page 10, the application must be signed to become legally effective (see Fig. 13.23).

Page 11 is completed by the German Patent Office (see Fig. 13.24).

On page 12, the fees are calculated and the desired payment method is specified (see Fig. 13.25).

13.12.2 Online Application

With the online service DPMAdirektWeb, an international registration can be filed online (see Fig. 13.26).[19]

[19] DPMA, "https://direkt.dpma.de/dpmadirektweb/index.xhtml", retrieved on June 21, 2021.

MM2 (E), page 4

4. APPOINTMENT OF A REPRESENTATIVE[8]

(a) Name:

(b) Address:

(c) E-mail address[9]:

(d) Telephone number[10]:

5. BASIC APPLICATION OR BASIC REGISTRATION

Basic application number:		Date of the basic application (dd/mm/yyyy):	
Basic registration number:		Date of the basic registration (dd/mm/yyyy):	

6. PRIORITY CLAIMED

☐ The applicant claims the priority of the earlier filing mentioned below:

Office of earlier filing:

Number of earlier filing (if available):

Date of earlier filing (dd/mm/yyyy):

[8] You **must** indicate the name, address and e-mail address of the representative; otherwise, WIPO cannot record the appointment.

[9] When a representative is appointed, WIPO will send all communications concerning this international application and the resulting international registration **only** to the e-mail address of the representative, except for a few communications where the Regulations require that WIPO send a copy to the holder (see the Note for Filing Form MM2). The applicant and the representative must ensure that the e-mail address indicated here is accurate and kept up to date.

[10] Indicating a phone number is not required, but it will allow WIPO to reach your representative if needed.

MM2 (E) – May 2021

Fig. 13.17 IR trademark application form page 4 (WIPO)

MM2 (E), page 5

If the earlier filing does not relate to all the goods and services listed in item 10, indicate in the space provided below the goods and services to which it does relate:

☐ If several priorities are claimed above, check this box and use a continuation sheet giving the information required for each priority claimed.

7. THE MARK

(a) Place the reproduction of the mark, as it appears in the basic application or basic registration, in the square below.

(b) Where the reproduction in item (a) is in black and white and color is claimed in item 8, place a color reproduction of the mark in the square below.

(c) ☐ The applicant declares that the mark is to be considered as a mark in standard characters.

(d) ☐ The mark consists exclusively of a color or a combination of colors as such, without any figurative element.

MM2 (E) – May 2021

Fig. 13.18 IR trademark application form page 5 (WIPO)

MM2 (E), page 6

8. COLOR(S) CLAIMED

(a) ☐ **The applicant claims color as a distinctive feature of the mark.**
Color or combination of colors claimed:

(b) **Indication, for each color, of the principal parts of the mark that are in that color** (as may be required by certain designated Contracting Parties):

9. MISCELLANEOUS INDICATIONS

(a) ☐ **Transliteration of the mark** (this information is compulsory where the mark consists of or contains matter in characters other than Latin characters, or numerals other than Arabic or Roman numerals):

(b) **Translation of the mark** (as may be required by certain designated Contracting Parties; **do not** check the box in item (c) if you provide a translation in this item):

(i) into English:

(ii) into French:

(iii) into Spanish:

(c) ☐ **The words contained in the mark have no meaning** (and therefore cannot be translated; **do not** check this box if you have provided a translation in item (b)).

(d) **Where applicable, check the relevant box(es) below:**

☐ Three-dimensional mark

☐ Sound mark

☐ Collective mark, certification mark, or guarantee mark

MM2 (E) – May 2021

Fig. 13.19 IR trademark application form page 6 (WIPO)

MM2 (E), page 7

(e) **Description of the mark** (as may be required by certain designated Contracting Parties, such as, for example, the United States of America)

(i) Description of the mark contained in the basic application or basic registration, where applicable (**only use this item** if the Office of origin requires to include this description in the international application for the purposes of item 13(a)(ii) of this form):

(ii) Voluntary description of the mark (any description of the mark by words, including the description contained in the basic application or registration, if you were not required to provide this description in item (e)(i) above):

(f) **Verbal elements of the mark (where applicable):**

(g) **The applicant wishes to disclaim protection for the following element(s) of the mark:**

MM2 (E) – May 2021

Fig. 13.20 IR trademark application form page 7 (WIPO)

MM2 (E), page 8

10. GOODS AND SERVICES[11]

(a) **Indicate below the goods and services for which the international registration is sought**[12]**:**

Class:	Goods and Services[13]:

(b) ☐ **The applicant wishes to <u>limit</u> the list of goods and services in respect of one or more designated Contracting Parties, as follows:**

Contracting Party:	Class(es) or list of goods and services for which protection is sought in this Contracting Party:

☐ If the space provided is not sufficient, check the box and use a **continuation sheet**.

[11] You can use the Madrid Goods and Services Manager (MGS) to find indications accepted by WIPO. In MGS, you can also find acceptance information for selected Contracting Parties. MGS is available at www.wipo.int/mgs.

[12] Use font "Courier New" or "Times New Roman", size 12 pt., or larger.

[13] Use semicolon (;) to separate indications or goods or services listed in a given class. For example:
09 Screens for photoengraving; computers.
35 Advertising; compilation of statistics; commercial information agencies.

MM2 (E) – May 2021

Fig. 13.21 IR trademark application form page 8 (WIPO)

MM2 (E), page 9

11. DESIGNATIONS[14]

Check the corresponding boxes:

- ☐ **AF** Afghanistan
- ☐ **AG** Antigua and Barbuda
- ☐ **AL** Albania
- ☐ **AM** Armenia
- ☐ **AT** Austria
- ☐ **AU** Australia
- ☐ **AZ** Azerbaijan
- ☐ **BA** Bosnia and Herzegovina
- ☐ **BG** Bulgaria
- ☐ **BH** Bahrain
- ☐ **BN** Brunei Darussalam[b]
- ☐ **BQ** Bonaire, Saint Eustatius and Saba[f,g]
- ☐ **BR** Brazil[e,h]
- ☐ **BT** Bhutan
- ☐ **BW** Botswana
- ☐ **BX** Benelux[i]
- ☐ **BY** Belarus
- ☐ **CA** Canada
- ☐ **CH** Switzerland
- ☐ **CN** China
- ☐ **CO** Colombia
- ☐ **CU** Cuba[e]
- ☐ **CW** Curaçao[f]
- ☐ **CY** Cyprus
- ☐ **CZ** Czech Republic
- ☐ **DE** Germany
- ☐ **DK** Denmark
- ☐ **DZ** Algeria
- ☐ **EE** Estonia
- ☐ **EG** Egypt
- ☐ **EM** European Union[a]
- ☐ **ES** Spain
- ☐ **FI** Finland
- ☐ **FR** France
- ☐ **GB** United Kingdom[b,j]
- ☐ **GE** Georgia
- ☐ **GG** Guernsey[b,k]
- ☐ **GH** Ghana
- ☐ **GM** Gambia
- ☐ **GR** Greece
- ☐ **HR** Croatia
- ☐ **HU** Hungary
- ☐ **ID** Indonesia
- ☐ **IE** Ireland[b]
- ☐ **IL** Israel
- ☐ **IN** India[b]
- ☐ **IR** Iran (Islamic Republic of)
- ☐ **IS** Iceland
- ☐ **IT** Italy
- ☐ **JP** Japan[e]
- ☐ **KE** Kenya
- ☐ **KG** Kyrgyzstan
- ☐ **KH** Cambodia
- ☐ **KP** Democratic People's Republic of Korea
- ☐ **KR** Republic of Korea
- ☐ **KZ** Kazakhstan
- ☐ **LA** Lao People's Democratic Republic
- ☐ **LI** Liechtenstein
- ☐ **LR** Liberia
- ☐ **LS** Lesotho[b]
- ☐ **LT** Lithuania
- ☐ **LV** Latvia
- ☐ **MA** Morocco
- ☐ **MC** Monaco
- ☐ **MD** Republic of Moldova
- ☐ **ME** Montenegro
- ☐ **MG** Madagascar
- ☐ **MK** North Macedonia
- ☐ **MN** Mongolia
- ☐ **MW** Malawi[b]
- ☐ **MX** Mexico
- ☐ **MY** Malaysia[b]
- ☐ **MZ** Mozambique[b]
- ☐ **NA** Namibia
- ☐ **NO** Norway
- ☐ **NZ** New Zealand[b]
- ☐ **OA** African Intellectual Property Organization (OAPI)[c]
- ☐ **OM** Oman
- ☐ **PH** Philippines
- ☐ **PK** Pakistan[b]
- ☐ **PL** Poland
- ☐ **PT** Portugal
- ☐ **RO** Romania
- ☐ **RS** Serbia
- ☐ **RU** Russian Federation
- ☐ **RW** Rwanda
- ☐ **SD** Sudan
- ☐ **SE** Sweden
- ☐ **SG** Singapore[b]
- ☐ **SI** Slovenia
- ☐ **SK** Slovakia
- ☐ **SL** Sierra Leone
- ☐ **SM** San Marino
- ☐ **ST** Sao Tome and Principe
- ☐ **SX** Sint Maarten (Dutch part)[f]
- ☐ **SY** Syrian Arab Republic
- ☐ **SZ** Eswatini
- ☐ **TH** Thailand
- ☐ **TJ** Tajikistan
- ☐ **TM** Turkmenistan
- ☐ **TN** Tunisia
- ☐ **TR** Turkey
- ☐ **TT** Trinidad and Tobago[b]
- ☐ **UA** Ukraine
- ☐ **US** United States of America[d]
- ☐ **UZ** Uzbekistan
- ☐ **VN** Viet Nam
- ☐ **WS** Samoa
- ☐ **ZM** Zambia
- ☐ **ZW** Zimbabwe

[14] You can find information on the procedures in national or regional offices in the Member Profile Database, available at www.wipo.int/madrid/memberprofiles.

MM2 (E) – May 2021

Fig. 13.22 IR trademark application form page 9 (WIPO)

MM2 (E), page 10

[a] The designation of the **European Union** covers its Member States (Austria, Belgium, Bulgaria, Croatia, Cyprus, Czech Republic, Denmark, Estonia, Finland, France, Germany, Greece, Hungary, Ireland, Italy, Latvia, Lithuania, Luxembourg, Malta, The Netherlands, Poland, Portugal, Romania, Slovakia, Slovenia, Spain, Sweden).

If the **European Union** is designated, it is compulsory to indicate a second language before the Office of the European Union, among the following (check one box only):

☐ French ☐ German ☐ Italian ☐ Spanish

Moreover, if the applicant wishes to claim the **seniority** of an earlier mark registered in, or for, a Member State of the European Union, the **official form MM17 must be annexed** to the present international application.

[b] By designating **Brunei Darussalam**, **Guernsey**, **India**, **Ireland**, **Lesotho**, **Malawi**, **Malaysia**, **Mozambique**, **New Zealand**, **Pakistan**, **Singapore**, **Trinidad and Tobago** or the **United Kingdom**, the applicant declares that he/she has the intention that the mark will be used by him/her or with his/her consent in that country in connection with the goods and services identified in this application.

[c] The designation of the African Intellectual Property Organization (**OAPI**) covers the following Member States: Benin, Burkina Faso, Cameroon, Central African Republic, Chad, Comoros, Congo, Côte d'Ivoire, Equatorial Guinea, Gabon, Guinea, Guinea-Bissau, Mali, Mauritania, Niger, Senegal, Togo.

[d] If the **United States of America** is designated, it is **compulsory to annex** to the present international application the official form (**MM18**) containing the declaration of intention to use the mark required by this Contracting Party. Item 2(e) of the present form should also be completed.

[e] **Cuba**, **Brazil** and **Japan** have made a notification under Rule 34(3)(a) of the Regulations. Their respective **individual fees are payable in two parts**. Therefore, if **Cuba**, **Brazil** or **Japan** is designated, only the first part of the applicable individual fee is payable at the time of filing the present international application. The second part will have to be paid only if the Office of the Contracting Party concerned is satisfied that the mark which is the subject of the international registration qualifies for protection. The date by which the second part must be paid, and the amount due, will be notified to the holder of the international registration at a later stage.

[f] Territorial entity previously part of the former Netherlands Antilles.

[g] Protection in **BQ** (Bonaire, Saint Eustatius and Saba) is granted automatically with the designation (see Information Notice No. 27/2011).

[h] By designating **Brazil**, the applicant declares that the applicant, or a company controlled by the applicant, effectively and lawfully conducts business in connection with the goods and services for which Brazil is being designated.

[i] The designation of **Benelux** covers the following States: Belgium, Luxembourg and the Netherlands.

[j] The designation of the **United Kingdom** covers England, Wales, Scotland, Northern Ireland, the British Overseas Territory of the Falkland Islands (Malvinas) and Gibraltar, as well as the two British Crown Dependencies of the Isle of Man and Jersey (see Information Notices No. 38/2015 and 77/2020).

[k] The Bailiwick of **Guernsey** is a self-governing British Crown Dependency (see Information Notice No. 77/2020).

12. SIGNATURE OF THE APPLICANT AND/OR THEIR REPRESENTATIVE

If required or allowed by the Office of origin.

By signing this form, I declare that I am entitled to sign it under the applicable law.
Signature:

MM2 (E) – May 2021

Fig. 13.23 IR trademark application form page 10 (WIPO)

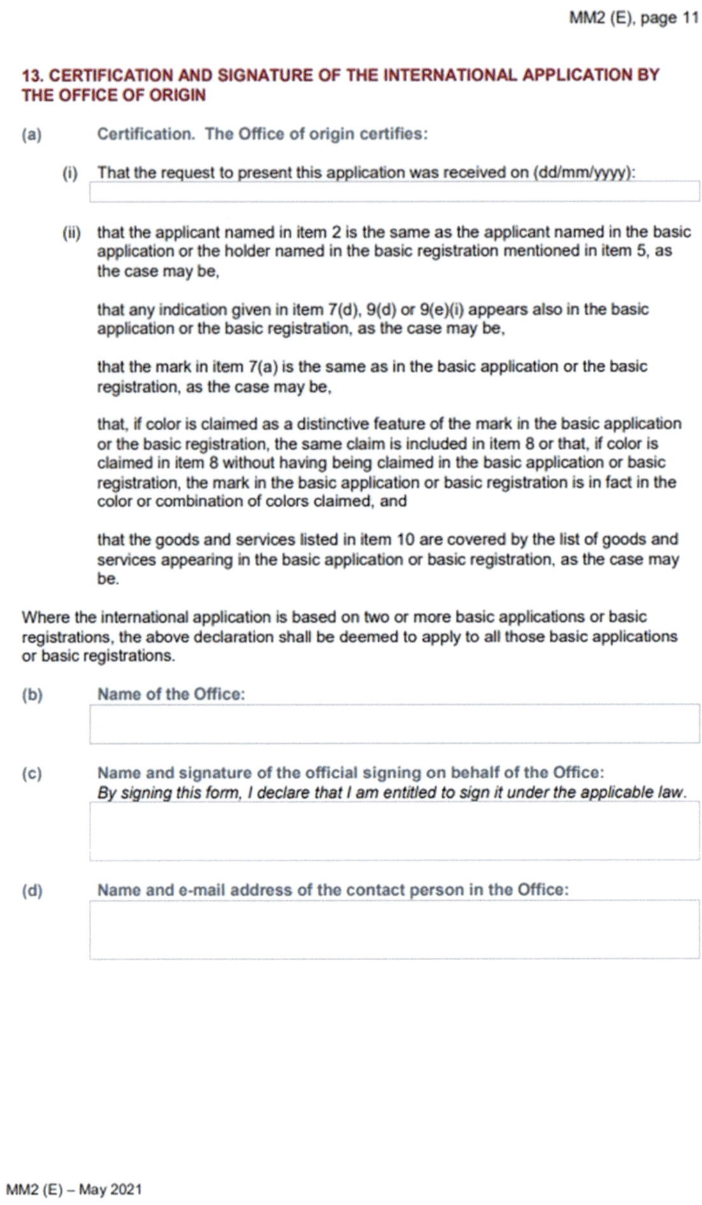

MM2 (E), page 11

13. CERTIFICATION AND SIGNATURE OF THE INTERNATIONAL APPLICATION BY THE OFFICE OF ORIGIN

(a) **Certification. The Office of origin certifies:**

(i) That the request to present this application was received on (dd/mm/yyyy):

(ii) that the applicant named in item 2 is the same as the applicant named in the basic application or the holder named in the basic registration mentioned in item 5, as the case may be,

that any indication given in item 7(d), 9(d) or 9(e)(i) appears also in the basic application or the basic registration, as the case may be,

that the mark in item 7(a) is the same as in the basic application or the basic registration, as the case may be,

that, if color is claimed as a distinctive feature of the mark in the basic application or the basic registration, the same claim is included in item 8 or that, if color is claimed in item 8 without having being claimed in the basic application or basic registration, the mark in the basic application or basic registration is in fact in the color or combination of colors claimed, and

that the goods and services listed in item 10 are covered by the list of goods and services appearing in the basic application or basic registration, as the case may be.

Where the international application is based on two or more basic applications or basic registrations, the above declaration shall be deemed to apply to all those basic applications or basic registrations.

(b) **Name of the Office:**

(c) **Name and signature of the official signing on behalf of the Office:**
By signing this form, I declare that I am entitled to sign it under the applicable law.

(d) **Name and e-mail address of the contact person in the Office:**

MM2 (E) – May 2021

Fig. 13.24 IR trademark application form page 11 (WIPO)

MM2 (E), page 12

FEE CALCULATION SHEET

(a) INSTRUCTIONS TO DEBIT FROM A CURRENT ACCOUNT

☐ The International Bureau is hereby instructed to debit the required amount of fees from a current account opened with the International Bureau (if this box is checked, it is not necessary to complete (b)).

Holder of the account:	
Account number:	
Identity of the party giving the instructions:	

(b) AMOUNT OF FEES (see Fee Calculator: www.wipo.int/madrid/en/fees/calculator.jsp)

Basic fee: 653 Swiss francs if the reproduction of the mark is in black and white only and 903 Swiss francs if there is a reproduction in color. (*For international applications filed by applicants whose country of origin is a Least Developed Country, in accordance with the list established by the United Nations (www.wipo.int/ldcs/en/country), 65 Swiss francs if the reproduction is in black and white only and 90 Swiss francs if there is a reproduction in color.*)	

Complementary and supplementary fees:

Number of designations for which complementary fee is applicable		Complementary fee	Total amount of the complementary fees		
	x	100 Swiss francs	=	=	
Number of classes of goods and services beyond three		Supplementary fee	Total amount of the supplementary fees		
	x	100 Swiss francs	=	=	

Individual fees (Swiss francs)[15]:

Designated Contracting Parties	Individual fee	Designated Contracting Parties	Individual fee

Total individual fees	=	
GRAND TOTAL (Swiss francs)	=	

(c) METHOD OF PAYMENT

Identity of the party effecting the payment:	

Payment received and acknowledged by WIPO	☐	WIPO receipt number	
Payment made to WIPO bank account IBAN No. CH51 0483 5048 7080 8100 0 Crédit Suisse, CH-1211 Geneva 70 Swift/BIC: CRESCHZZ80A	☐	Payment identification	dd/mm/yyyy
Payment made to WIPO postal account (within Europe only) IBAN No. CH03 0900 0000 1200 5000 8 Swift/BIC: POFICHBE	☐	Payment identification	dd/mm/yyyy

[15] Where individual fees have been declared, you will pay these fees instead of the standard fees **except** where the designated Contracting Party and the Contracting Party of the holder are both States bound by the Protocol and the Agreement, in which case, a complementary fee is payable.

MM2 (E) – May 2021

Fig. 13.25 IR trademark application form page 12 (WIPO)

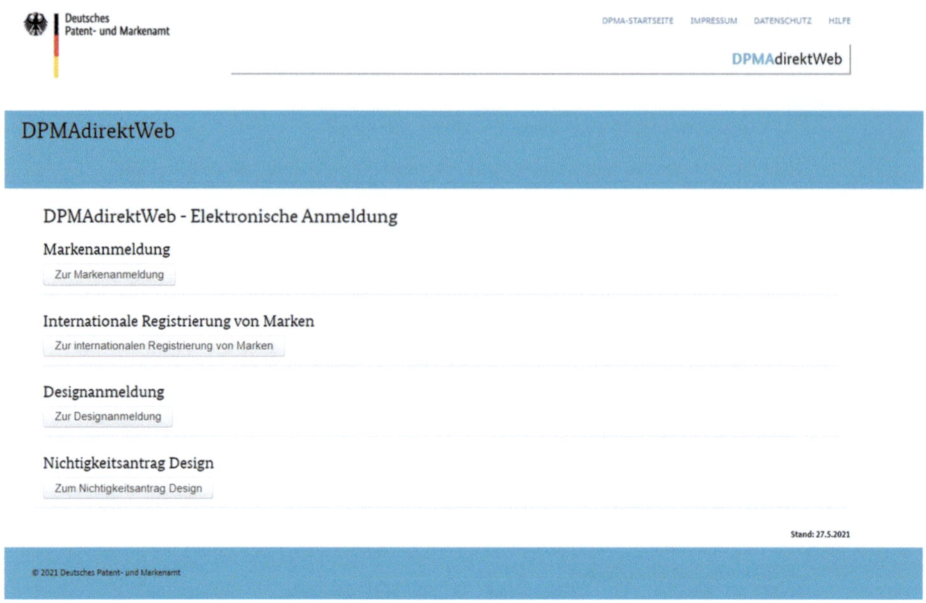

Fig. 13.26 IR trademark with DPMAdirektWeb (DPMA)

14 Opposition to a Trademark

An opposition is a cost-effective procedure for removing a trademark from a register. It is also the simplest way to challenge a trademark. However, the opposition period must be observed. An opposition can be filed against a newly registered or to-be-registered trademark within a 3-month period. For a German trademark, the 3-month period begins after publication of the registration of the trademark in the register.[1] An opposition can be based on an earlier trademark or an earlier company sign.[2] The opposition fee must be paid within the opposition period.[3]

Example

Best Software GmbH has registered the trademark "Pasquale" for software in the trademark register of the German Patent Office. Best Software GmbH has set up trademark monitoring and has discovered that Bad Software GmbH has registered the trademark "Pasqualo" for class 9 for software. Best Software GmbH can now file an opposition against the registration of the trademark by Bad Software GmbH.◂

An opposition can only be filed by the owner of an earlier trademark or an earlier company sign.[4] The opposition must be substantiated.[5] This means that the opponent

[1] Section 42 (1) sentence 1 of the Trademark Act.

[2] Section 42 (2) of the Trademark Act or Article 46 (1) of the European Union Trademark Regulation.

[3] Article 46 (3) sentence 2 of the European Union Trademark Regulation.

[4] Section 42 (1) sentence 1 of the Trademark Act.

[5] Article 46 (3) sentence 1 of the European Union Trademark Regulation.

T. H. Meitinger, *Trademark Without a Lawyer*,
https://doi.org/10.1007/978-3-662-73127-7_14

must explain why there is a likelihood of confusion between the younger trademark and their own. It is possible to request that the grounds for opposition be submitted within an extension period of two months. This request is usually granted. The owner of the earlier trademark must demonstrate that there is an identity of signs and goods and services[6] or that there is a similarity of the signs and the goods and services[7]. A special case arises with a well-known trademark in Germany. In this special case, only the similarity of the signs needs to be proven.[8] The opposition is forwarded to the owner of the younger trademark. They are usually given a period of four months to respond to the opposition.

The opposition procedure typically involves extensive correspondence between the parties and the patent office, as the grounds for opposition are submitted, the owner of the younger trademark responds to the grounds, and possibly a plea of non-use is raised—i.e., the owner of the earlier trademark is requested to prove genuine use of their trademark—and the owner of the younger trademark responds to the evidence of use. Approximately four months are granted for each response. An opposition procedure can therefore take between one and two years at first instance.

14.1 Opposition Against a German Trademark

After publication of the registration of a trademark, any holder of an earlier right may file an opposition against the registration of the trademark with the German Patent Office. The opposition must be based on the likelihood of confusion between the two trademarks. The opposition must be filed within three months after publication of the registration of the trademark. Within this period, the opposition fee of € 250 must also be paid. If the opposition is based on additional opposition trademarks of the same owner, an additional € 50 must be paid for each further opposition trademark.[9] An opposition against a German trademark can be filed using form W7202.[10]

On the first page of form W7202 (see Fig. 14.1), field (1) is for the details of the owner of the challenged trademark. In field (2), the details of the opponent are to be entered.

On the second page of the opposition form (see Fig. 14.2) of the German Patent Office, field (4) indicates whether the opposition is directed against the entire trademark or only certain goods and services. In field (5), the preferred method of payment, especially for the opposition fee, is to be specified. In addition, the opposition must be signed in field (7).

[6] Section 9 (1) no. 1 of the Trademark Act.

[7] Section 9 (1) no. 2 of the Trademark Act.

[8] Section 9 (1) no. 3 of the Trademark Act.

[9] DPMA, "https://www.dpma.de/marken/widerspruch_loeschung/index.html", accessed on 4.10.2021.

[10] DPMA, "https://www.dpma.de/docs/formulare/marken/w7202.pdf", accessed on June 17, 2021.

Deutsches
Patent- und Markenamt

Deutsches Patent- und Markenamt
Markenabteilungen
80297 München

W7202 4.19 1

(1) **Registernummer/Aktenzeichen der Marke, gegen deren Eintragung, bzw. Nummer der international registrierten Marke, gegen deren Schutzerstreckung auf Deutschland sich der Widerspruch richtet**

Name, Vorname/Firma des Inhabers der angegriffenen Marke

Widerspruch gegen

- **die Eintragung einer Marke**
- **die Schutzerstreckung einer international registrierten Marke auf Deutschland**

3

☐ per Telefax **vorab** am TT MM JJJJ
an Telefaxnummer +49 89 2195 - 4000

(2) **Widersprechender**

☐ Weitere/r Widersprechende/r siehe Anlage *(bitte ein separates Blatt DIN A4 bzw. einen Datenträger verwenden)*

Name, Vorname/Firma *(ggf. einschließlich Rechtsform entsprechend registerrechtlicher Eintragung)*

Straße, Hausnummer *(kein Postfach)*

Postleitzahl **Ort** **Land** *(nur bei ausländischen Adressen)*

Telefonnummer **E-Mail-Adresse**

Telefaxnummer **Geschäftszeichen**

(3) **Sendungen**
des Deutschen Patent- und Markenamts sind zu richten an den

☐ **Vertreter des Widersprechenden** *(Rechts- oder Patentanwalt, Patentassessor)*

oder

☐ **Zustellungsbevollmächtigten des Widersprechenden**

Name, Vorname/Bezeichnung

Straße, Hausnummer

Postleitzahl **Ort** **Land** *(nur bei ausländischen Adressen)*

Telefonnummer **E-Mail-Adresse**

Telefaxnummer **Geschäftszeichen**

ggf. Nummer der Allgemeinen Vollmacht

Fig. 14.1 Opposition against German trademark page 1 (DPMA)

W 7 2 0 2 4 . 1 9 2

(4) Der Widerspruch richtet sich gegen

☐ alle Waren/Dienstleistungen der angegriffenen Marke

☐ folgende Waren/Dienstleistungen der angegriffenen Marke

☐ siehe Anlage *(bitte ein separates Blatt DIN A4 bzw. einen Datenträger verwenden)*

(5) Gebührenzahlung in Höhe von ________ **€**

Zahlung per Banküberweisung

☐ **Überweisung**
(dreimonatige Zahlungsfrist beachten)

Zahlungsempfänger:
Bundeskasse Halle/DPMA
IBAN: DE84 7000 0000 0070 0010 54
BIC (SWIFT-Code): MARKDEF1700

Anschrift der Bank:
Bundesbankfiliale München
Leopoldstr. 234, 80807 München

Zahlung mittels SEPA-Basis-Lastschrift

☐ Ein gültiges **SEPA-Basis-Lastschriftmandat** *(Formular A 9530)*

☐ liegt dem DPMA bereits vor *(Mandat für mehrmalige Zahlungen)*

☐ ist beigefügt

☐ **Angaben zum Verwendungszweck** *(Formular A 9532)* des Mandats mit Mandatsreferenznummer sind beigefügt

! Bitte beachten Sie die Kostenhinweise auf der letzten Seite dieses Formulars.

(6) Anlagen

☐ ___ **Formular(e) W 7202.1** *(Registrierte Widerspruchkennszeichen, auf die sich der Widerspruch stützt)*

☐ ___ **Formular(e) W 7202.2** *(Nicht registrierte Widerspruchskennzeichen, auf die sich der Widerspruch stützt)*

Mindestens eines dieser Formulare muss beigefügt werden!

☐ **Verzeichnis der Waren und/oder Dienstleistungen**, gegen die sich der Widerspruch richtet - Anlage Feld (4)

☐ **Vollmacht**

☐ **Doppelstücke sämtlicher Widerspruchsunterlagen** (§ 17 Abs. 2 DPMAV)

☐ ________________

(7) Unterschrift

Der Unterschrift ist der Name in Druckbuchstaben oder Maschinenschrift hinzuzufügen; bei Firmen die Bezeichnung laut registerrechtlicher Eintragung mit Angabe der Stellung/Funktion des Unterzeichnenden.

Bitte beachten Sie hinsichtlich der Verarbeitung Ihrer personenbezogenen Daten unser Merkblatt A 9106 „Datenschutz bei Schutzrechtsanmeldungen". Dieses finden Sie unter www.dpma.de: Service – Formulare – Sonstige Formulare – Hinweise zum Datenschutz.

Datum	Unterschrift/en, ggf. Firmenstempel	Funktion/en der/des Unterzeichner/s

Fig. 14.2 Opposition against German trademark page 2 (DPMA)

If the decision on an opposition is made by a senior civil servant and one disagrees with the decision, a so-called "reminder" (Erinnerung) can be filed.[11] A senior civil servant has not studied law. A reminder results in a further review of the decision by an official at the Patent Office who has studied law. As a rule, a reminder does not lead to a change in the decision, as the second examiner usually confirms the decision of the first examiner. The only remaining option is an appeal to the Federal Patent Court.[12]

14.2 Opposition Against a European Union Trademark

Within a period of three months after publication of a trademark application, an opposition can be filed against the forthcoming registration of a European Union trademark.[13] The owner of a European Union trademark, a German trademark, or an international registration designating Germany or the EU[14] may file an opposition on the grounds of likelihood of confusion with their own trademark.

The opposition must be submitted in writing and substantiated. In addition, an opposition fee must be paid.[15] When filing the opposition, the opponent may request that the grounds for opposition be submitted within an additional period of two months after expiry of the opposition period, which is usually granted.[16]

14.3 Defense Against Opposition

A defense against an opposition particularly involves demonstrating that there is no likelihood of confusion in terms of sound, visual appearance, or conceptual meaning. In addition to arguing against the merits of the opposition, one should consider raising the plea of non-use and launching a counterattack by initiating cancellation proceedings against the earlier trademark.

14.3.1 Plea of Non-use

The owner of the younger trademark may raise a plea of non-use. In this case, the owner of the earlier trademark must prove that the earlier trademark has been used within the last five years prior to the filing or priority date of the younger trademark. Proof is only

---

[11] Section 64 (1) sentence 1 of the Trademark Act.

[12] Section 66 (1) sentence 1 of the Trademark Act.

[13] Article 46 (1) of the European Union Trademark Regulation.

[14] EM is the abbreviation for a European Union trademark in an international registration.

[15] Article 46 (3) of the European Union Trademark Regulation.

[16] Article 46 (4) of the European Union Trademark Regulation.

required if the earlier trademark has already been registered for five years at that time.[17] The evidence should always be provided with a statutory declaration and extensive documentation (offers, advertising materials, advertising expenses, invoices, etc.).[18] If evidence of use is provided only for part of the registered goods and services, the trademark is at least subject to cancellation for the remaining goods and services.

If the earlier trademark has already been registered for more than five years, the plea of non-use should always be raised. The opponent must then submit documents proving that their trademark has been used in the last five years. The documents must therefore be dated for the correct period, must show the earlier trademark, and the trademark must have been used for the goods and services for which it is registered. These requirements should be checked for each individual sheet of the submitted documents.

► **Tip** If your own trademark is attacked by an opposition based on a trademark that has been registered for more than five years, the plea of non-use should be raised as a matter of course.

14.3.2 Counterattack: Cancellation Proceedings

A counterattack can be launched with cancellation proceedings. Cancellation proceedings are particularly advisable if the opponent has difficulties proving use of their trademark. In this case, a cancellation action for revocation is recommended. Another advantage of cancellation proceedings is that the opposition procedure is suspended as long as the cancellation proceedings are pending. This puts pressure on the opponent to seek an out-of-office or out-of-court settlement. A cancellation action also puts the entire trademark of the opponent at risk, thereby increasing the pressure on the opponent.

► **Tip** By initiating invalidity proceedings before the Patent Office or cancellation proceedings before a civil court against the opposition trademark, a willingness to negotiate on the part of the owner of the earlier trademark can be created.

---

[17] Section 43 (1) sentence 1 of the Trademark Act or Article 47 (2) sentence 1 of the European Union Trademark Regulation. For a German trademark, it should be noted that the five-year period begins at the end of the opposition period or the end of an opposition procedure.

[18] Section 43 (1) sentence 2 of the Trademark Act.

14.4 Differences Between German and European Opposition Proceedings

In an opposition against a German trademark, the opposition is filed against an already registered trademark. This means that a German trademark can be enforced despite a pending opposition.[19] The opposition period for a European Union trademark begins after publication of the trademark application by the EUIPO.[20]

In the European procedure, an opposition is filed against the registration of a trademark in the trademark register. The opposed trademark is therefore still a trademark application and not a registered trademark. An opposition can be filed with the EUIPO within three months after publication of the trademark application.[21] Registration of the European Union trademark only takes place after completion of the opposition procedure. While the opposition is pending before the EUIPO, a European Union trademark cannot be enforced. Therefore, as the attacked party, it is advisable to prolong a German opposition procedure and to expedite a European opposition procedure as much as possible.

The European opposition procedure is highly regulated. There are only a few opportunities to submit statements to the EUIPO. Therefore, in opposition proceedings before the EUIPO, care should be taken to ensure that your own statements fully and comprehensively reflect your position.

The German opposition procedure is less strictly structured, so it can be assumed that a large number of statements can be submitted to the Patent Office. The exchange of statements typically ends when the parties jointly request a decision from the Patent Office. Alternatively, the Trademark Office informs the parties that sufficient submissions have been made and that a decision will now be issued. The German opposition procedure can take up to three years.

A special feature is that the prevailing party in a European opposition procedure is awarded € 300. Of course, this amount of € 300 is disproportionate to the actual costs incurred. For patent attorney support in an opposition procedure, costs between € 2000 and € 3000 should be expected.

14.5 Alternative to Opposition: Coexistence Agreement

An opposition procedure is associated with high costs for both parties. It is therefore advisable, before initiating an opposition procedure or shortly thereafter[22], to attempt to achieve a delimitation by restricting the goods and services of the younger trademark so that both trademarks can coexist.

[19] Section 42 (1) sentence 1 of the Trademark Act.

[20] Article 46 (1) of the European Union Trademark Regulation.

[21] Article 46 (1) of the European Union Trademark Regulation.

[22] Section 42 (4) of the Trademark Act.

15 Cancellation of a Trademark

If the opposition period has expired, a trademark can only be removed from the trademark register through a cancellation procedure. For this purpose, a nullity procedure can be conducted before the German Patent Office. Alternatively, a cancellation action can be brought before a civil court.

15.1 Withdrawal of the Application

A trademark is deleted from the trademark register if the trademark owner withdraws the trademark.[1]

15.2 Cancellation Due to Absolute Grounds

A trademark is deleted from the trademark register upon request if there were absolute grounds for refusal to its registration.[2] In this case, the trademark is not eligible for protection, the applicant could not be the owner of a trademark, or absolute grounds for refusal prevent the trademark from remaining in the register. The application must be submitted in writing, stating the grounds for cancellation as well as the facts and evidence.[3] In addition, a fee of € 400 must be paid within three months of filing the application.

[1] Section 39 (1) Trademark Act.

[2] Section 8 Trademark Act.

[3] Section 53 (1) sentences 1 and 2 Trademark Act.

T. H. Meitinger, *Trademark Without a Lawyer*,
https://doi.org/10.1007/978-3-662-73127-7_15

The two most important reasons for cancellation due to violation of absolute grounds for refusal are the lack of distinctiveness of the trademark and the infringement of the need to keep certain terms free for the goods and services for which the trademark was registered. A lack of distinctiveness exists if the relevant public does not recognize that a trademark is present at all.

Example

Best Software GmbH labels its products with the sign "Super" as an indication of origin so that the relevant public knows that the product comes from Best Software GmbH. However, the sign "Super" will not be able to fulfill this function. The relevant public will simply perceive "Super" as advertising and not as an indication of origin. The sign "Super" lacks distinctiveness.◄

Example

Best Software GmbH labels its products with "Software for financial accounting." Such a sign is descriptive and therefore cannot be registered as a trademark. If this sign was mistakenly registered as a trademark for class 9 for software, it is subject to cancellation and can be removed from the register through a cancellation procedure due to the absolute ground for refusal of infringing the need to keep the term free.◄

A German trademark can only be cancelled due to absolute grounds for refusal within 10 years. After 10 years, cancellation on the basis of absolute grounds for refusal is excluded.[4]

15.2.1 A Trademark Becomes a Generic Term

A trademark that has become a generic term can be deleted from the registers of the patent offices. Cancellation of a trademark that has become a generic term is relatively simple at the EUIPO. For this, an application for cancellation of the trademark due to absolute grounds for refusal, namely the description of the goods and services, is filed.

When applying for cancellation of a generic term at the German Patent Office, it must also be stated that the trademark owner has not made any efforts to prevent the development of their trademark into a generic term.

A trademark becomes a generic term when it is so dominant in a market that the trademark is equated with the market itself. In this case, the relevant public assumes that the trademark describes the market. A generic term thus loses its ability to indicate the origin of the marked products. If a trademark has become a generic term, competitors

[4] Section 50 (2) sentence 3 Trademark Act.

need the trademark to describe the goods or services. The generic term describes the characteristics of a product or service. Examples of trademarks that have become generic terms are “Fön” for hair dryer, “Googeln” for internet search, “Post” for delivery services, and “Tempo” for paper tissues.

To counteract genericization, the trademark owner can run advertisements to prevent the relevant public from subsuming competing products under the trademark. In particular, the trademark owner is allowed to request that a publisher of reference works refrain from using the trademark generically. The publisher can be asked to include a note with the trademark indicating that the designation is a registered trademark and not a generic term.

15.3 Cancellation Due to Prior Rights

The owner of an earlier right can file an application for cancellation of a trademark with the German Patent Office. The cancellation application must be submitted in writing. In addition, the facts and evidence on which the application is based must be stated.[5] An application for cancellation can only be filed by the owner of the earlier trademark. The trademark owner may base their application for cancellation on several trademarks.

An application for cancellation is unsuccessful if forfeiture has occurred or if the earlier trademark was already subject to cancellation for non-use at the time of the filing or priority date of the later trademark.[6] Furthermore, a cancellation procedure is not admissible if the earlier trademark is void due to absolute grounds for refusal. Absolute grounds for refusal exist in particular if the trademark lacks distinctiveness, infringes the need to keep certain terms free, was filed in bad faith, is misleading, violates public order or accepted principles of morality, or contains official test or certification marks, state emblems, or national flags.

The application for cancellation of a German trademark is filed with the German Patent Office. A fee of € 400 must be paid within three months. If the application is based on two or more earlier trademarks, a fee of € 100 must be paid for each additional earlier trademark.

The German Patent Office forwards the cancellation application to the owner of the later trademark. If the trademark owner objects to the cancellation of their trademark, the official procedure before the Patent Office begins. If the trademark owner does not object within two months, their trademark is removed from the trademark register.

As an alternative to a nullity procedure before the German Patent Office, an action can be brought before a civil court.

[5] Section 53 (1) sentences 1 and 2 Trademark Act.

[6] Section 26 Trademark Act.

15.4 Cancellation Due to Lapse

A trademark can be removed from the trademark register upon request if it has not been genuinely used within a period of five years.[7] A cancellation procedure due to lapse can be conducted as an official procedure before the German Patent Office or as an action before a civil court.

The application to initiate a nullity procedure before the Patent Office must be submitted in writing to the Patent Office and substantiated with the facts and evidence.[8] In addition, a fee of € 100 must be paid within a period of three months.

The Patent Office notifies the trademark owner of the application for cancellation due to lapse. The trademark owner may object to the application within a period of two months. Otherwise, the trademark is removed from the trademark register. If the trademark owner files an objection, the applicant is required to pay a continuation fee of € 300. As soon as the applicant has paid the continuation fee, the cancellation procedure before the German Patent Office begins. If the continuation fee is not paid, the cancellation procedure will not be initiated.

Cancellation of a trademark due to lapse for non-use is a common reason for a cancellation procedure. Non-use may occur because the trademark is no longer used at all or because it is used in a modified form. Non-use despite use (in a modified form) exists if the two forms of the trademark have a different distinctive character. The distinctive character of two trademarks is different if, for the relevant public, they are two different trademarks. For word marks, a different spelling does not change their distinctive character. The graphic design is irrelevant for a word mark. The problem of a changed distinctive character mainly arises with figurative marks and word/figurative marks.

Non-use exists if, in the five years preceding the filing of the cancellation application or the initiation of legal action, and in the five years preceding the final decision on the application or action, no use has taken place. Resumption of use in the four months prior to the application for cancellation or initiation of legal action is disregarded if the trademark owner was made aware of the impending lapse of their trademark.

The trademark owner can avert cancellation of their trademark by providing proof of use. Use should always be demonstrated with a statutory declaration and materials such as advertising flyers, catalog entries, promotional materials, offers, and invoices. Expenditures for advertising, market research, etc. also help to prove use. These documents must show the protected sign in connection with the goods and services. In addition, there must be a date indicating that use took place during the relevant periods.

---

[7] Section 49 Trademark Act.

[8] Section 53 (1) sentences 1 and 2 Trademark Act.

15.5 Cancellation Procedure Before a Civil Court

A cancellation procedure can be initiated by filing an action before a civil court[9]. The alternative to legal action before a civil court is the official nullity procedure before the Patent Office. The official nullity procedure before the Patent Office is inexpensive but slow. The cancellation procedure before the Regional Court is fast but carries a high cost risk in the event of losing.

15.6 Cancellation of a European Union Trademark

A cancellation procedure for a European Union trademark takes place as an official procedure before the EUIPO.

15.7 Cancellation of an International Registration

The part of an internationally registered trademark extended to Germany can be challenged with a cancellation procedure before the German Patent Office or with a cancellation action before an ordinary German court.[10]

15.8 Costs of a Cancellation Procedure

The costs of an official cancellation procedure are approximately € 4000 in the first instance.

For the costs of a judicial cancellation procedure, according to the Lawyers' Remuneration Act (RVG)[11] a total risk of at least € 20,000 to € 40,000, i.e., costs for the lawyers of both parties and court costs, must be expected.

[9] The court of first instance is a Regional Court.

[10] Section 115 Trademark Act.

[11] RVG: Lawyers' Remuneration Act.

16 Trademark Monitoring

A trademark search is recommended before registering a trademark and at regular intervals after the trademark has been registered, preferably on a monthly basis. Conducting a search prior to trademark registration can prevent the registration from conflicting with an earlier right. Regular searches after trademark registration serve trademark monitoring and prevent the "dilution" of one's own trademark rights. In particular, this prevents identical or similar trademarks from becoming established for the same or similar goods and services, which could damage the image of one's own trademark and prevent additional sales.

A trademark owner should regularly search for more recent trademarks. Third-party, more recent trademarks that could be confused with one's own trademark can cause damage in the market. It may happen that the owner of a more recent trademark, by offering inferior quality goods and services, damages the image of the older, confusable trademark. To prevent this sometimes irreparable damage, confusable, more recent trademarks should be deleted or measures should be taken to ensure that their goods and services are so clearly distinguished from those of the older trademark that any risk of confusion is excluded.

To eliminate a disruptive, more recent trademark, the opposition procedure is provided. This allows for a cancellation procedure to be conducted cost-effectively and quickly. However, the three-month opposition period must be observed. An opposition can only be effectively filed within three months from the date of publication of the trademark.[1] It is therefore advisable, for example, to conduct a search for confusable, more recent trademarks once a month in order not to miss the opposition period. The opposition

[1] Section 42(1) of the Trademark Act or Article 46(1) of the EU Trademark Regulation.

T. H. Meitinger, *Trademark Without a Lawyer*,
https://doi.org/10.1007/978-3-662-73127-7_16

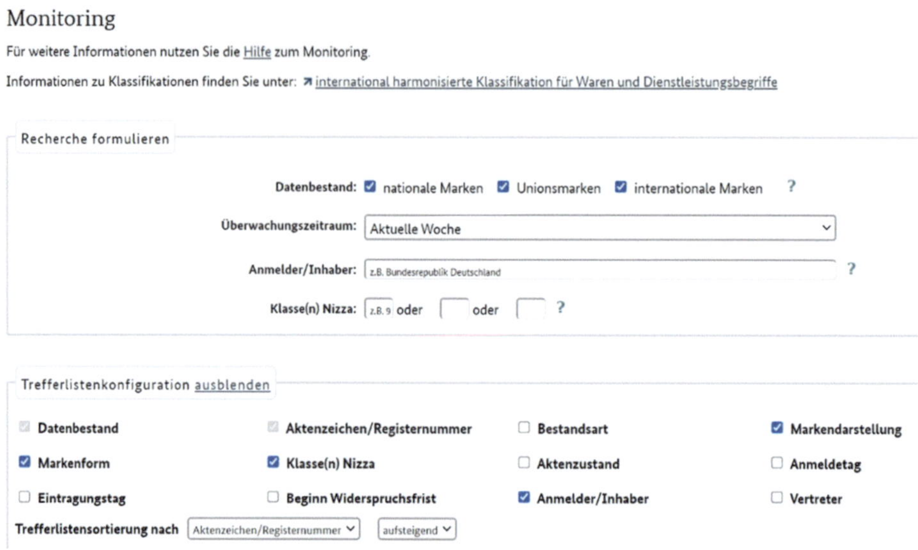

Fig. 16.1 Monitoring by the German Patent Office

procedure takes place before the patent office, for example before the German Patent Office or the EUIPO.

If the owner of a trademark discovers through trademark monitoring that a third party has registered or is using a similar or identical trademark for similar or identical goods and services, the owner of the older trademark must take action against this more recent trademark in order not to risk or restrict their own rights. This risk arises from the forfeiture of rights from the trademark.

16.1 DPMAregister

In the online service DPMAregister[2] of the German Patent Office, trademark monitoring can be conducted under the link "https://register.dpma.de/DPMAregister/marke/monitoring"[3] (see Fig. 16.1).

With the monitoring tool of the German Patent Office, searches can be conducted by owner, by Nice classes, and by owner in combination with Nice classes. Monitoring can be carried out for German trademarks, EU trademarks, and international registrations

[2] DPMAregister is a combination of DPMA (German Patent and Trademark Office) and register.

[3] DPMA, "https://register.dpma.de/DPMAregister/marke/monitoring", accessed on June 19, 2021.

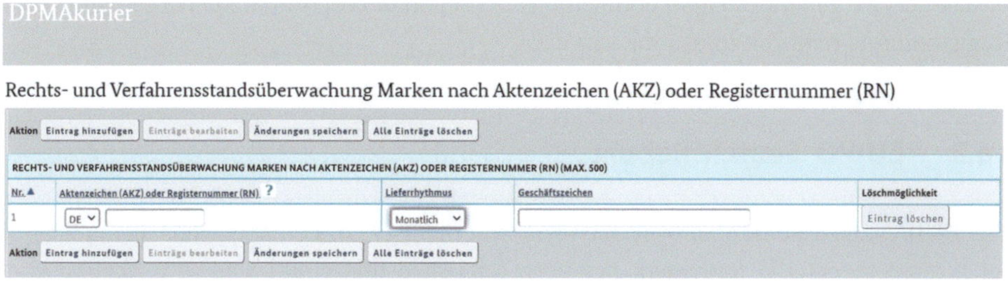

Fig. 16.2 DPMAkurier: Monitoring by register number (DPMA)

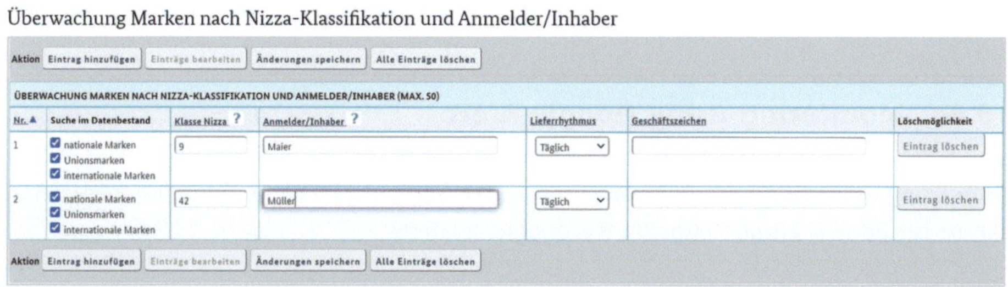

Fig. 16.3 DPMAkurier: Monitoring by classes and owners (DPMA)

designating Germany or EM[4]. Truncation of entries is possible. There is no distinction between upper and lower case.

16.2 DPMAkurier

Search results from the DPMAregister service can be monitored with the online tool DPMAkurier, so that notifications are sent by email in the event of legal or procedural changes (see Fig. 16.2).[5]

However, with the DPMAkurier service of the German Patent Office, independent monitoring can also be set up for German trademarks, EU trademarks, and international registrations designating Germany and EM[6].[7] Automatic monitoring can be set up according to two criteria, namely by Nice classes, by owners, and by Nice classes in combination with owners (see Fig. 16.3).

[4] EM is the abbreviation for an EU trademark in an international registration.

[5] DPMA, "https://register.dpma.de/DPMAregister/kurier/markenakz", accessed on June 19, 2021.

[6] EM is the abbreviation for an EU trademark in an international registration.

[7] DPMA, "https://register.dpma.de/DPMAregister/kurier/uebersicht", accessed on June 19, 2021.

The service generates email notifications of potentially relevant trademark applications. Registration is required to use the service.

16.3 EUIPO Search Report

The EUIPO automatically informs the owners of older EU trademarks when an application for a more recent EU trademark has been published that could lead to a likelihood of confusion. The owner of the older trademark can use the three-month opposition period to file an opposition with the EUIPO.

Therefore, an owner of an EU trademark can limit their trademark monitoring to the respective national patent offices. Additional monitoring of EU trademarks is not required.

16.4 Monitoring using Google Alert

With Google Alert, the use of trademarks on the internet can be monitored. Google Alert can be set up at the link "https://www.google.de/alerts".

17 Enforcing a Trademark

A trademark constitutes an economic monopoly and grants the trademark owner a right to injunctive relief against unauthorized use of their trademark. The trademark owner must enforce this right themselves. Enforcement can be pursued through a request for justification, a warning letter (cease-and-desist letter), a preliminary injunction, or by filing a lawsuit.

17.1 Request for Justification

A request for justification is an extrajudicial request for a statement. Especially when the facts of the case are unclear, a request for justification is advisable to avoid unnecessary risks. Such a request inquires about circumstances that may hinder the enforcement of claims. The recipient of the request is not obliged to respond. Typically, the recipient will reply and attempt to demonstrate that enforcement of the claim has no prospect of success. Unlike a warning letter, a request for justification does not contain a threat of legal proceedings.

A request for justification sets a deadline within which the recipient is to explain why they believe they are entitled to use the trademark. In contrast to a warning letter, a request for justification does not include a declaration to cease and desist or an undertaking. No legal action is threatened in a request for justification.

A request for justification is sensible if it can be assumed that the trademark infringement was accidental. This is particularly the case with small businesses that do not have a professionally established monitoring system.

T. H. Meitinger, *Trademark Without a Lawyer*,
https://doi.org/10.1007/978-3-662-73127-7_17

Example

The company Best Software GmbH discovers that Worst Software GmbH is also labeling its software with the trademark "Pasquale." Best Software GmbH writes to Worst Software GmbH: "We have noticed that you are also distributing your software under the trademark 'Pasquale.' Please explain to us, within a period of 2 weeks, why you believe you are entitled to do so."◄

17.2 Warning Letter (Cease-and-Desist Letter)

If it can be assumed that the infringer acted knowingly or if a request for justification yields no result, a warning letter is an appropriate means to counter trademark infringement. A typical value in dispute for a trademark warning letter ranges between 50,000 and 250,000 €.

Example

The company Best Software GmbH finds that Worst Software GmbH continues to label its software as "Pasquale." Furthermore, Worst Software GmbH has simply ignored the request for justification.◄

17.2.1 Active Legitimation

A warning letter is not permissible if it is not issued by the trademark owner. For example, if the owner of a company issues a warning letter, even though the company itself, and not its owner, is the trademark owner, the warning letter is ineffective. In this case, the owner of the company is not entitled to issue a warning letter. The party issuing the warning letter lacks active legitimation.

On the other hand, the owner of the company could still have active legitimation if they have received an exclusive license to use the trademark from the company. This would need to be proven.

17.2.2 Passive Legitimation

The party issuing the warning letter must address the correct party, i.e., the actual trademark infringer. Otherwise, the warning letter is also ineffective. If a GmbH or UG is the infringer, it is advisable to obtain the correct company name and address from the commercial register.

17.2.3 Claims Are Too Broadly Formulated

It would be incorrect to issue a warning letter to a trademark infringer for goods and services for which the trademark is registered, but which the infringer has neither produced nor offered. The party issuing the warning letter can only demand that the infringer refrain from offering those products and services under the trademark that they have already unlawfully produced or offered. Therefore, the entire list of goods and services for the trademark should not be included in the warning letter without verification. For all goods and services not offered or produced by the recipient, the warning letter is unjustified.

17.2.4 Procedure for a Warning Letter

A warning letter serves to enforce a claim for injunctive relief out of court. A warning letter is typically prepared by a patent attorney or lawyer on behalf of a trademark owner and sent to the infringer. The infringer is given a deadline within which the trademark infringement must cease. In addition, a trademark warning letter always includes a cease-and-desist declaration subject to penalty, which the infringer must sign. Only a cease-and-desist declaration subject to penalty can eliminate the risk of repeat infringement.

The contractual penalty in a cease-and-desist declaration is usually set above € 5000 so that, in the event of legal proceedings, the regional court (Landgericht) and not the local court (Amtsgericht) has jurisdiction.[1]

Furthermore, a warning letter demands information on the extent to which the trademark has been used and the revenue generated with the trademark, in order to determine damages. The warning letter typically also requires reimbursement of the attorney's fees incurred by the party issuing the warning letter.

Upon receipt of a warning letter, a patent attorney or lawyer with relevant trademark expertise should be consulted as soon as possible, since the deadlines set for the recipient are typically short.

An experienced patent attorney will first check whether the warning letter is justified. If this is the case, they will attempt to mitigate the consequences for their client, for example by negotiating a sell-off period for goods labeled with the trademark. Other points that can be negotiated even in the case of a justified warning letter include waiving the obligation to provide information or reducing the costs for the recipient.

Beforehand, it should be clarified, for example in a telephone conversation with the opposing attorney, what the intention behind the warning letter is. If an injunction is

[1] Section 23 No. 1 of the Courts Constitution Act (Gerichtsverfassungsgesetz).

intended and the warning letter is justified, the cease-and-desist declaration should be signed and a new trademark sought. Otherwise, it may be possible to negotiate a license agreement with the party issuing the warning letter.

If the warning letter is formally incorrect, for example if the recipient's designation is not exact, the cease-and-desist declaration should still be signed in the case of a substantively justified warning letter, since the party issuing the warning letter can of course immediately issue a correct warning letter. In addition, legal proceedings may already be imminent.

17.3 Negative Declaratory Action

A negative declaratory action can provide legal certainty as to whether a trademark infringement exists. This can dispel any growing uncertainty, for example among the business partners of the alleged infringer. Therefore, if, as the recipient of a warning letter, you are convinced that no trademark infringement exists and the party issuing the warning letter does not promptly initiate legal proceedings, you should consider seeking judicial clarification yourself by filing a negative declaratory action.

17.4 Preliminary Injunction

After a warning letter, a preliminary injunction can be sought or a lawsuit can be filed with the regional court. The prerequisite for a preliminary injunction is urgency. If the trademark infringement occurred too long ago, a preliminary injunction is no longer possible.

A preliminary injunction can only be used to enforce a claim for injunctive relief. If injunctive relief is not the objective, a preliminary injunction is not the appropriate means to enforce one's interests.

A preliminary injunction can be issued unilaterally, i.e., without a hearing of the opposing party, or in a bilateral procedure. In the latter case, the court will promptly schedule an oral hearing to reach a decision. Currently, there is a trend toward deciding on preliminary injunctions in bilateral rather than unilateral proceedings.

A protective brief can be filed to counter a threatened preliminary injunction. In a protective brief, one's own arguments can be explained, especially why one's own trademark does not create a likelihood of confusion with an earlier trademark. By filing a protective brief, the potentially affected party can increase the likelihood of a bilateral procedure in the event of an application for a preliminary injunction. The protective brief is filed in the central protective brief register in Frankfurt. If an application for a preliminary injunction is submitted to a judge, they will first check the protective brief register to see if a protective brief has been filed. The arguments in the protective brief will be considered in the decision on the preliminary injunction or an oral hearing will be scheduled at short notice. Filing a protective brief is always advisable if you have already received a warning letter and believe that the warning letter is unjustified or even abusive.

A preliminary injunction is sent to the opposing party, who must immediately cease the infringing activity. The preliminary injunction only decides on injunctive relief. This is followed by the main proceedings, in which the trademark owner's further claims can be asserted.

17.5 Action Before an Ordinary Court

Immediate filing of a lawsuit is generally not advisable. At the very least, a warning letter should be sent to the infringer beforehand. Otherwise, the infringer can immediately acknowledge the claim, and the trademark owner will have to bear all costs, i.e., the costs of both parties and the court costs.[2] For this reason, at least a warning letter with a short deadline should be sent to the opposing party beforehand.

Prompt filing of a lawsuit may be necessary if limitation is imminent. It should be noted that a warning letter does not suspend the statute of limitations. In the event of impending limitation, a warning letter with a short deadline or immediate filing of a lawsuit may therefore be required.

The lawsuit must specify why the defendant has infringed the trademark, i.e., why a trademark infringement exists. It must be clearly set out why there is a likelihood of confusion between the infringer's trademark and the earlier trademark.

A claim for information can be asserted in the lawsuit. The defendant must then specify exactly to whom and at what price the infringing product was sold. In addition, they must, if applicable, state from whom they obtained the infringing product. Based on the infringer's information, damages can be calculated.

Legal proceedings take approximately one year. This means that after filing the lawsuit, an oral hearing can be expected to be scheduled within a year.

In legal proceedings, the principle of losing party pays applies with regard to costs. This means that the losing party must bear all costs, i.e., the legal fees of both parties and the court costs. The total cost risk of trademark infringement proceedings before a regional court is approximately between 10,000 and 30,000 €.

[2] Section 93 of the Code of Civil Procedure (ZPO); this legal situation is referred to as a "Klageüberfall" (ambush lawsuit).

18 Obligation to Use a Trademark

A trademark can represent a significant intervention in economic life. The legislator does not want such interventions to be possible with trademarks that are not in use. For this reason, trademark law provides for an obligation to use the trademark.

A trademark owner must use their trademark. Otherwise, the trademark becomes subject to cancellation. This means that the trademark can be challenged at any time by a third party and removed from the register. If no challenge occurs, the trademark remains in the register. If the trademark owner resumes use of the trademark, the risk of cancellation is remedied.

Use that preserves rights exists if the trademark is seriously used by its owner in the territory of Germany for the goods and services for which the trademark is registered.[1] Serious use is present if significant sales are generated with products marked by the trademark. At the very least, appropriate marketing measures must be demonstrated. If proof of serious use is required, in addition to suitable documentation, a statutory declaration should always be submitted. Use by a licensee or a third party with the consent of the trademark owner is considered use by the trademark owner.[2]

Example

Best Software GmbH develops a new product and secures the trademark "Pasquale" for software. The development of the product is delayed by two years. It also makes no sense to carry out marketing measures at this stage. Therefore, the trademark cannot be used in the first two years. Does the trademark lapse?◄

---

[1] Section 26(1) of the Trademark Act.

[2] Section 26(2) of the Trademark Act.

T. H. Meitinger, *Trademark Without a Lawyer*,
https://doi.org/10.1007/978-3-662-73127-7_18

For the owner of a newly registered trademark, there is initially no problem. Within the first five years, a trademark does not lapse due to non-use. However, the trademark owner must have commenced use of their trademark within this non-use grace period. The term "non-use grace period" is therefore misleading, as it is a period within which use must begin at the latest. The term "use commencement period" would be more accurate.

The grace period for use begins on the day after no further opposition can be filed against the trademark or an opposition proceeding has been concluded, and ends five years later.[3]

Example

Best Software GmbH is not making progress with its product, which is to bear the "Pasquale" trademark. After almost five years, alarm bells are ringing. It is foreseeable that the product can only be launched on the market shortly after the end of the grace period. What should Best Software GmbH do to prevent the lapse of its "Pasquale" trademark?◄

It can indeed happen that the development of a complex technical product takes years, or that the development of another product was prioritized. However, the grace period for use is a statutory period and cannot be extended. A trademark owner must therefore at least begin to use the trademark within these five years.

Example

At a trade fair, Best Software GmbH learns through contacts that its direct competitor, Bad Software GmbH, has become aware of the "Pasquale" trademark. A cancellation action may be planned to delay the launch of Best Software GmbH's innovative product. Furthermore, Bad Software GmbH also appears to be working on a similar product that is likewise not yet market-ready. What should Best Software GmbH do?◄

For use that preserves rights, it may be sufficient to at least prepare marketing measures. For example, if orders for printing advertising materials have been placed with a print shop shortly before the end of the grace period, the design of packaging bearing the printed trademark has been prepared, and an order for the creation of a web presence for the product with the trademark has been placed, this may constitute use that preserves rights. While this does not prove serious use, it does demonstrate serious preparation for use of the trademark. If actual use is no longer possible before the end of the grace period, the risk of cancellation can still be averted by preparing to use the trademark.

[3] Section 49(1) sentence 1 of the Trademark Act.

However, serious preparation for use should only be regarded as a last resort. In cancellation proceedings, serious preparation is usually more difficult to prove than actual use. In practice, proof of serious preparation fails more often than proof of use.

Example

Best Software GmbH has failed to use its "Pasquale" trademark for software before the end of the grace period. Bad Software GmbH files a request for cancellation due to lapse with the Patent Office, and the Patent Office requests a statement from Best Software GmbH. Best Software GmbH is alarmed and immediately begins extensive use of its trademark.◄

The trademark owner cannot wait until a request for surrender or a request for cancellation due to lapse is filed with the Patent Office. Once a request for surrender is received or a request for cancellation due to lapse has been filed with the Patent Office, commencing use can no longer remedy the lapse of the trademark.

18.1 Grace Period for Use

Within the first five years after registration, a trademark does not have to be used and can still be enforced.[4] The term "grace period for use" is misleading, because the trademark must nevertheless be seriously used within the first five years after registration. Otherwise, the trademark becomes subject to cancellation immediately after the five-year grace period expires. If no cancellation action is brought or no invalidity proceedings due to lapse are initiated, the risk of cancellation of a trademark can be remedied by commencing serious use.

The grace period for use begins with the registration of the trademark in the register. However, if an opposition has been filed against the trademark, the five-year period starts with the conclusion of the opposition proceedings.[5]

18.2 Intervening Rights

An interruption of use for less than five years is harmless. However, if the trademark is not used for more than five years, there is a risk that a cancellation proceeding will succeed or so-called intervening rights will arise.

Intervening rights are younger trademarks that conflict with an older trademark due to a likelihood of confusion and that were filed during a period when the older trademark

[4] Section 25(1) of the Trademark Act.

[5] Section 26(5) of the Trademark Act.

was subject to cancellation. These intervening rights cannot be challenged by the older trademark, even if the risk of cancellation of the older trademark has since been remedied. The owner of the older trademark must accept the existence of the younger, confusingly similar trademark.

18.3 Caution: Trademark in the USA

In the USA, there is no five-year grace period for use after registration of a trademark. The trademark applicant must provide proof of use immediately after filing. If a trademark is extended to the USA through international registration, the applicant is granted a period of several years to provide proof of use. When renewing a US trademark, proof of use must be provided. If this proof is not successful, the trademark will be deleted by the USPTO[6].

[6] USPTO: United States Patent and Trademark Office.

19 Facelifting and Relaunch

After several years of use, it may become necessary to "refresh" a trademark. Alternatively, it may be necessary to adapt a trademark to a realignment of the company. It is possible that a figurative mark or a word/figurative mark contains elements that no longer reflect the company's self-image.

19.1 Use Preserving Rights

When undertaking a facelift or relaunch, it is particularly important to ensure use preserving rights. Use preserving rights are maintained as long as the facelift does not result in a deviation from the previous mark that alters the distinctive character of the trademark.[1]

In the case of a word mark, it can almost always be assumed that a facelift does not lead to a change in the distinctive character of the mark. The reason for this is that a word mark provides protection for all fonts and typefaces.

For a word/figurative mark or a figurative mark, a facelift very quickly leads to a change in the distinctive character. From a trademark law perspective, a facelift of a word/figurative mark or a figurative mark often means the creation of a new mark that has nothing in common with the registered mark.

[1] Section 26 paragraph 3 of the Trademark Act.

T. H. Meitinger, *Trademark Without a Lawyer*,
https://doi.org/10.1007/978-3-662-73127-7_19

19.2 Forfeiture Due to Non-Use of the Trademark

A facelift and a relaunch carry the risk that a new mark with a new distinctive character will result. In this case, the previously registered mark becomes subject to cancellation due to forfeiture after five years. On the other hand, the new mark does not enjoy protection through the previous entry in the trademark register.

19.3 Word/Figurative Mark

A word/figurative mark consists of a text and an image component. However, the text component does not have the same scope of protection as a word mark, which covers all fonts and typefaces. Therefore, a word/figurative mark should be regarded as a figurative mark when assessing use preserving rights.

19.4 Figurative Mark

A figurative mark can, for example, be a logo. The facelift can in particular result in the altered mark no longer being confusable with the old mark. In this case, there are definitely two new marks. The recognition of the old mark can no longer be utilized by the company. On the other hand, if the old mark is taken over by another company, that company can benefit from the recognition of the mark. If this company introduces inferior goods to the market under the old mark, reputational damage may occur. In addition, the marketing expenses for the old mark are lost. The new mark must be established with new marketing expenditures. Care should be taken to ensure that, as long as the old mark can still be enforced, so-called intermediate rights are challenged.

19.5 Recommended Method

A "modernization" of a word mark is generally not a problem. The facelift of an antiquated-looking figurative mark or word/figurative mark must be carried out within a framework that does not result in a different distinctive character compared to the mark originally registered in the trademark register. In case of doubt, a new figurative mark or word/figurative mark should be registered, and both the older and the newer mark should be maintained simultaneously.

In this case, two marks coexist in the same country for a certain period. The older mark can eventually be abandoned by not paying the renewal fees.

20 Value of a Trademark

The monetary value of a brand is relevant when purchasing the brand. The purchase of the brand may be associated with the sale or acquisition of a company that owns the brand. Furthermore, the purpose of the valuation may be to determine how marketing measures affect the value of a brand.

There is no generally accepted method for calculating the monetary value of a brand. The various calculation models can only provide guidance to avoid a gross misjudgment of brand value. Figure 20.1 presents three classic methods of brand valuation.

20.1 Incurred Costs as Brand Value

The value of a brand can be considered as the sum of the costs incurred. This includes the costs of trademark registration, costs for maintenance, brand monitoring, and extension to additional countries. The costs of a brand can be regarded as the lower limit of the brand's value. This is particularly true since, as a rule, only minor costs are incurred for a brand.

20.2 Brand Income Value

In an income-based calculation, the focus is on the revenues generated. In particular, the revenues from licenses for the use of the brand by third parties, whether actually or potentially achieved, are considered.

T. H. Meitinger, *Trademark Without a Lawyer*,
https://doi.org/10.1007/978-3-662-73127-7_20

Fig. 20.1 Classic methods of brand valuation

Traditional methods of brand valuation
• Cost of a brand (lower value) • Income value of a brand (income from license payments, for example) • Price premium method (what premium does the brand generate)

20.3 Price Premium Model

In a price premium model, the premium that can be demanded by the market because a product is branded is considered.

Example

Best Software GmbH sells its products in a first market, where its brand "Pasquale" is regarded as a well-known and respected mark, and in a second market, where Best Software GmbH has only recently started operating and its brand is therefore unknown. In the first market, Best Software GmbH can achieve a price of € 2500 for its financial accounting software. In the second market, the product can only be sold for € 2000. The brand is therefore responsible for enabling a premium of € 500 in the first market.◄

20.4 Value for a Company

The value of a brand results from the economic monopoly and from its recognition value.

Example

Best Software GmbH has been able to establish recognition and general acceptance of the quality of its products with its brand "Pasquale." The relevant market participants, especially potential customers, are therefore more likely to make a purchase when a software product is branded "Pasquale." It is therefore important that Best Software GmbH ensures the market remains "clean," particularly by enforcing the economic monopoly of the brand.◄

20.5 Criteria for Brand Valuation

Common criteria for the economic valuation of a brand are presented in practice.

20.5.1 Relevance in the Target Market

A brand is relevant in a market if the relevant market participants, especially the respective customers, are familiar with the brand. A brand that has not achieved relevance in its target market is worthless. In this case, it must be weighed whether the brand has potential and should be developed, or whether it should be abandoned.

20.5.2 Relative Strength

Relative strength means that a brand enjoys higher acceptance among the relevant market participants compared to those of the competition. A brand that does not exhibit relative strength compared to relevant competitors is worthless.

20.5.3 Sales with the Brand

If no relevant sales are generated with products branded with the mark, the brand is not valuable.

21 Using a Third-Party Trademark

It may not even be necessary to register and build your own brand. Perhaps an existing third-party brand can be used.

21.1 Explicit Permission

The use of a third party's trademark can be permitted by a license agreement. The license for a German trademark does not have to cover the entire territory of Germany, but can, for example, be limited to Bavaria or Hesse or even just individual regions or cities. Typically, a license is limited to five or ten years. A license can be a simple or an exclusive license. A simple license exists if, in principle, there can be several licensees for the same territory. An exclusive license exists if the licensee alone is entitled to use the trademark.

21.2 Exhaustion of Trademark Rights

Exhaustion of trademark rights for certain goods occurs if, within the European Economic Area, goods have already been marked with the trademark with the consent of the trademark owner. For example, if products with a trademark are lawfully acquired in Spain and the trademark owner holds the trademark not in Spain but in Germany, the purchaser of the goods from Spain can still offer and sell these products in Germany. The German trademark cannot be enforced against them.

T. H. Meitinger, *Trademark Without a Lawyer*,
https://doi.org/10.1007/978-3-662-73127-7_21

21.3 Forfeiture of Trademark Rights

Forfeiture occurs if the owner of a trademark has known for more than five years about the use of their trademark by a third party and has not taken action against it. After forfeiture, enforcement of the trademark is no longer possible.[1]

However, it applies that each sale of a product restarts the five-year period. In a landmark decision, the Hard Rock Cafe chain sued the Hard Rock Cafe Heidelberg, which did not belong to the chain. The Federal Court of Justice ruled that the Hard Rock Cafe Heidelberg was allowed to continue its business of catering to guests due to forfeiture. However, it was prohibited from selling T-shirts with the Hard Rock Cafe chain's logo. The Federal Court of Justice determined that each individual sale of a T-shirt restarts the five-year forfeiture period.[2]

Forfeiture of an ongoing service, such as catering to guests, is therefore still possible. The Federal Court of Justice considers an ongoing service to be a continuous use, which allows only a single start of the forfeiture period. In contrast, forfeiture of the sale of goods, for example of merchandise products, is effectively excluded by the Federal Court of Justice's decision, since each sale must be considered individually for forfeiture. This means that the trademark owner forfeits their rights for a product only after five years from its sale, and forfeiture must be recalculated for each individual product. For goods, therefore, forfeiture can be considered ineffective.

A prerequisite for forfeiture is that knowledge of the trademark infringement exists. In a company, the question arises as to which person must specifically have knowledge for the forfeiture period to begin. For example, it could be that only the head of the patent department was aware. Is this sufficient, or should the managing director also have knowledge of the trademark infringement? As a rule, it is assumed that a head of a significant department of the company must have positive knowledge of the trademark infringement for the forfeiture period to begin. Therefore, if the head of marketing or the head of the patent department becomes aware of the trademark infringement, the forfeiture period is deemed to have started. If a production employee becomes aware of the trademark infringement, this will not be sufficient.

Positive knowledge is presumed if the trademark owner has failed to become aware of the trademark infringement due to gross negligence. For example, if a company has a well-staffed trademark department, it is reasonable to expect the company to become aware of a newer trademark. This applies in particular if this could have been easily achieved through standard trademark monitoring.

Positive knowledge is also assumed, for example, if the trademark owner and the infringer know each other as competitors and, in particular, the infringing trademark was used at the infringer's trade fair booth.

---

[1] Section 51 paragraph 2 sentence 1 of the Trademark Act.

[2] Federal Court of Justice, 15.8.2013, I ZR 188/11, Gewerblicher Rechtsschutz und Urheberrecht, 2013, p. 1161—Hard Rock Cafe.

21.4 Own Older Trademark

If you yourself have your own registered, older trademark that is likely to be confused with a third-party, younger trademark, and the older trademark is not subject to cancellation, the third-party, younger trademark can be used, as the third-party, younger trademark is subject to cancellation. It is possible to have the third-party, younger trademark cancelled on the basis of the older, own trademark.

21.5 Descriptive Use of Trademarks and Spare Parts Business

A dealer must be able to indicate from which manufacturer the products they offer originate. A trademark can therefore be used to describe a product. For example, the use of the description "Brake discs for use in all Mercedes-Benz vehicles manufactured from 2006 onwards" is permissible even for businesses that do not belong to Daimler AG. However, the use of a Mercedes-Benz star in the advertising brochure of an independent car repair shop, which gives the impression that the repair shop belongs to the Daimler Group, is not permitted. Indicating that the independent car repair shop specializes in the repair of Mercedes-Benz vehicles, however, is permissible.

21.6 Revocation Due to Non-Use of the Trademark

If a trademark has been registered in the register for more than five years and has not been used in a manner that preserves rights within the last five years, it becomes subject to cancellation and can no longer be enforced. This trademark can be used by any third party.

If an older trademark lapses, the trademark should be registered by oneself and a cancellation request should be filed against the older trademark. Otherwise, there is a risk that the owner of the older trademark will resume use and thereby cure the lapse.

22 Trademark Law and Amazon

Since May 2017, Amazon Corporation has offered the possibility to register a registered trademark. In doing so, Amazon has aligned itself with trademark law.

22.1 Amazon Brand Registry

With the brand registry introduced in May 2017[1], the online marketplace Amazon is attempting to combat brand counterfeiting and fraudulent activities involving hacked Amazon accounts. A prerequisite for brand registration is that the trademark owner provides detailed information about their brand. In addition, the trademark owner must upload images of the product bearing the trademark and present the product's packaging. Information about target customers, distribution channels, and manufacturing locations is also required. With this information, Amazon can identify counterfeit products and curb brand piracy. Furthermore, Amazon notifies the trademark owner of suspicious listings, and the owner can block fraudulent offers using their brand.

A prerequisite for Amazon brand registration is a registered trademark, such as a German trademark or a European Union trademark. Registration of unregistered trademarks, as was previously possible, is no longer allowed. Amazon grants the registered trademark owner greater control over the product title, product description, and product images.

Every product on Amazon's online marketplace receives an individual number, the so-called ASIN (Amazon Standard Identification Number). Using the ASIN, Amazon can identify sellers offering the same product and present their offers to customers.

[1] Amazon, "https://brandregistry.amazon.de", accessed on June 16, 2021.

T. H. Meitinger, *Trademark Without a Lawyer*,
https://doi.org/10.1007/978-3-662-73127-7_22

With Amazon brand registration, unauthorized listing under an ASIN is virtually excluded, as the trademark owner can control the sellers. The trademark owner will only authorize their licensees and distribution partners. The trademark owner can exclude other companies from so-called listing via the ASIN. As a result, Amazon registration can be used to exclude competitors from the marketplace. The trademark owner alone can determine who even has the chance to win the Buy Box. The Buy Box is the shopping cart field. It is a box on the right side, which, when clicked, adds the product to the shopping cart. Securing the Buy Box is of crucial importance for Amazon sellers.

22.2 Unlocking a Listing

If a listing has been blocked, action should be taken as quickly as possible to prevent further listings from being blocked, which could ultimately lead to the account being deleted. The reason for deletion should be checked, and if it is not justified—especially if there is no trademark infringement or counterfeiting—the person who filed the complaint with Amazon should be contacted. In the event of a block, it is not advisable to contact Amazon directly.

Experience shows that unlocking can be expedited if "pressure" is applied to the complainant. In particular, it should be examined whether their trademark can be removed from the trademark register through an opposition or cancellation action.

The person who filed the complaint should also be informed precisely which email address and which text should be sent to facilitate the unblocking. Amazon has a specific email address for complaints.

Managing a Trademark Portfolio 23

A company always has a brand with its company name. In addition to this brand, the company usually has one or more brands for individual products or product segments. Therefore, a company almost always has to manage a brand portfolio. Over the years, the brand portfolio typically increases. A sensible, consistent brand strategy is recommended to keep the costs of the brand portfolio under control.

It makes sense to conduct an evaluation of the individual brands in a portfolio. In this process, unsuccessful brands can be discontinued in order to use the freed-up resources to build promising brands.

23.1 Value of a Brand

To assess the brands in the brand portfolio, a valuation of the brands can be carried out. The value of the brands can be determined based on the criteria "incurred costs," "earnings value through license fees," "price premium due to product labeling with the brand," "relevance in the target market," "relative strength of the brand in the market," "sales with the brand," and possible "acceptance problems."[1]

[1] See Chap. 20 Value of a Trademark.

T. H. Meitinger, *Trademark Without a Lawyer*,
https://doi.org/10.1007/978-3-662-73127-7_23

23.2 Brands with Different Positionings

It can be useful to maintain several brands in the same market. If the brands have different positioning and each brand targets different subsegments of the market, the market can be better covered by the different brands.

23.3 Regional Brands or a Uniform Global Brand

It may be sensible to maintain different brands in different regions. On the other hand, it can be advantageous to build a global brand, possibly with descriptive additions that vary from region to region.

23.4 Cannibalization of Brands

If it turns out that two brands in the same market are targeting the same customer group and are therefore competing with each other, consideration should be given to discontinuing one of the two brands.

23.5 Mono-Brand Strategy (Branded House)

In a branded house brand architecture, there is a single dominant brand. This brand serves as an umbrella for all business areas of the company. Different products and services are labeled under this strong brand.

A branded house brand strategy is obvious for companies with complementary or conceptually related goods and services. An example of this is the British mobile communications company Vodafone Group, which offers only products and services in the field of telecommunications.

23.6 Multi-Brand Strategy (House of Brands)

The house of brands brand strategy advocates a separate brand for each product area. The company often does not appear in the foreground, so customers are often unaware that the respective products come from the same company.

Especially in the consumer goods sector, there are companies with a large number of different strong brands. An example of this is the British consumer goods group Unilever plc. (see Table 23.1).

Table 23.1 Brands of unilever plc

Food	Hygiene and Cosmetic Products	Cleaning and Detergent Products
Ben & Jerry's	Axe	Coral (formerly: Korall)
Bertolli	Dove	Domestos
Du Darfst	dusch das	OMO
Flora (Becel)	Lux	Viss
Knorr	Pond's	
Langnese (ice cream)		
Lipton		
Magnum (ice cream)		
Mazola		
Mondamin		
Pfanni		
Sanella		
Slim-Fast		

Another example of a house of brands brand strategy is shown by Procter & Gamble Company, whose brands have completely different positioning and identities (see Table 23.2).

Another important brand of Procter & Gamble Company is "Braun" for electrical appliances.

23.7 Hybrid Form

The brand policy of Bayer Aktiengesellschaft can be regarded as a hybrid form of branded house and house of brands. Bayer has strong brands, such as Aspirin or Bepanthen, and a strong umbrella brand, namely Bayer.

23.8 Optimization of a Brand Portfolio

Brands require effort and cost money. The deadlines for payment of renewal fees must be monitored and the fees must be paid. In addition, trademark monitoring must be established and confusingly similar younger brands from third parties must be opposed.

Table 23.2 Brands of procter & gamble company

Food	Hygiene and Cosmetic Products	Cleaning and Detergent Products
Femibion (dietary supplements)	Always (feminine hygiene products)	Ariel (detergent)
Vicks or Wick (cold remedies)	Alldays (panty liners)	Dash (detergent)
	Blend-a-dent (oral hygiene)	Fairy (dishwashing liquid)
	Blend-a-med (oral hygiene)	Febreze (fabric refresher)
	Blendax (oral hygiene)	Lenor (fabric softener)
	Charmin (toilet paper, in Germany since 2009 Zewa Soft Samtstark or Zewa Moll Deluxe)	Meister Proper (detergent and household cleaner)
	Gillette (wet razors)	Swiffer (floor cleaning system)
	Head & Shoulders (hair care)	
	Herbal Essences (hair care)	
	Luvs (diapers)	
	Oil of Olaz (cosmetics)	
	Oral-B (oral hygiene)	
	Pampers (diapers and wet wipes)	
	Pantene (hair care)	
	Tampax (tampons)	

Each brand in a brand portfolio should be examined to determine whether it still generates relevant sales. When reviewing a brand portfolio, one may come across brands that are no longer in use.

Example

Best Software GmbH started its business activities with the sale of humidifiers 20 years ago. In the meantime, no more humidifiers are manufactured or sold. Best Software GmbH still owns a "MyBestAir237" trademark in the United Kingdom and an "Aero237" trademark for humidifiers in Spain. Renewal fees should no longer be paid for these trademarks.◀

Individual national trademarks in EU member states may become obsolete through the registration of a European Union trademark. By claiming the seniority of the national trademarks, the trademarks can be consolidated into a European Union trademark while retaining the respective earlier priority date of the national trademarks. However, the respective goods and services must be identical for this. Alternatively, instead of many national trademarks, an international registration can be filed to obtain trademark protection. An international registration can subsequently be extended to additional countries.

24 Practical Tips

Important practical tips are provided to help avoid beginner mistakes.

24.1 Caution with the Use Grace Period

The term "use grace period" is misleading. Although, within the first five years after registration or after the expiry of the opposition period, no successful cancellation proceedings for non-use can be brought against a trademark, this becomes possible the day after this 5-year period. At that point, it is relevant whether the trademark has been used in a manner that preserves rights within the preceding five years.

This means that the trademark must be used during the use grace period to prevent it from being cancelled for non-use after the end of the grace period. It is therefore necessary that the trademark is already in use during the grace period in order to exclude cancellation for non-use. The term "use grace period" is misleading. A better designation for the first five years of a trademark would be "use commencement period."

Use must be in a manner that preserves rights. Use that preserves rights exists if the trademark owner has made genuine use of the trademark in the domestic market for the registered goods or services.[1] Use that preserves rights must occur for all goods and services for which the trademark is registered. Therefore, it makes no sense to register a trademark for goods and services for which it is foreseeable that the trademark will never be used.

[1] Section 26(1) of the Trademark Act or Article 18(1) of the EU Trademark Regulation.

T. H. Meitinger, *Trademark Without a Lawyer*,
https://doi.org/10.1007/978-3-662-73127-7_24

24.2 Trademark Snatched Away?

The trademark registers of the patent offices contain an endless variety of trademarks. It is therefore possible that a desired word mark is already reserved. In this case, instead of a word mark, a word/design mark could be registered with the patent office. The prerequisite for this is that the added design element is dominant over the text element. Simple embellishments or typical graphic elements that, for example, symbolize the text element, are generally not sufficient. Rather, independently distinctive design elements must be included, and the text element must be written in small letters. The attempt to file a word/design mark instead of a word mark is therefore generally not advisable, as the text element may only be given a subordinate representation. In this case, it makes sense to come up with a new word mark.

If a word mark fails because there are absolute grounds for refusal—i.e., the word mark is descriptive or consists of one or more words that, for example, are understood as promotional and not as an indication of origin—these grounds for refusal could also be overcome by strong design elements in a word/design mark. However, here too, it is a better alternative to find a new word mark.

24.3 Sailing Close to the Wind

It often happens that companies achieve great success with a trademark and competitors try to capitalize on that success. A concrete example was the battle over the golden Easter bunny with a red collar from Chocoladefabriken Lindt & Sprüngli AG, which is protected as a 3D trademark (see Fig. 24.1).

Hans Riegelein & Sohn GmbH & Co. KG in Cadolzburg also launched an Easter bunny in gold-colored packaging. Of course, without the Lindt inscription and without a red collar.

Lindt took legal action against Riegelein's Easter bunny. In several proceedings before regional and higher regional courts and two appearances before the Federal Court of Justice, Riegelein ultimately prevailed. The proceedings were correspondingly expensive for both sides. It can be assumed that each party incurred legal costs in the six-figure range. "Sailing close to the wind" can therefore be successful, but one should be aware that it can lead to fierce disputes with the trademark owner.

	Database	DB	?	EM
111/210	Brand number	RN/ AKZ	? ?	001698885
540	Brand presentation	MD	?	
	Wording of the trademark	MD	?	Lindt Gold Bunny
	Acquired distinctiveness			No
270	First language			German
270	Second language			English
550	Brand form	MF	?	Three-dimensional trademark
550	Trademark type EU trademarks	EUIPOMF	?	Three-dimensional trademark
591	Designation of colors	FA	?	Gold, red, brown.
551	Brand category	MK	?	Individual brand
220	Application date	AT	?	June 8, 2000
151	Date of entry in the register	ET	?	07/06/2001
730	Owner	INH	?	Chocoladefabriken Lindt & Sprüngli AG, 8802, Kilchberg, CH

Fig. 24.1 Register excerpt of the Lindt 3D bunny (DPMA)

24.4 Adventures of a Trademark Application in Saudi Arabia

With an international registration, a trademark can be registered in many countries around the world with a single application. An important exception is Saudi Arabia. Saudi Arabia is not a member state of the Madrid Agreement or the Protocol Relating to the Madrid Agreement. Therefore, only a national trademark application is possible in Saudi Arabia.

The process of registering a trademark in Saudi Arabia is extremely bureaucratic, lengthy, and expensive. Numerous documents must be prepared and certified. A trademark application in Saudi Arabia should therefore be carefully considered.

24.5 Caution: Secure Your Domain in Time

There are companies that search the trademark register for newly registered trademarks for which no domains have yet been reserved. These domains are then registered and offered for sale to the trademark applicant. Naturally, these companies expect to be paid for this "service." To avoid this, a search for available domain names should be conducted at the same time as the trademark application, and these should be reserved immediately.

24.6 Caution: Misleading Invoices

More and more often, trademark owners receive invoices from companies that appear official and request payment for services related to trademarks. It should be noted that patent offices generally do not send invoices, as it is up to the trademark owner to monitor due dates and pay the fees on time.

Before paying an invoice, the origin should therefore be carefully checked. The EUIPO has created a database of misleading invoices that can be used to identify such invoices. The database can be accessed at the link "https://euipo.europa.eu/ohimportal/de/misleading-invoices."[2]

[2] EUIPO, "https://euipo.europa.eu/ohimportal/de/misleading-invoices," accessed on June 18, 2021.

MIX
Papier aus verantwortungsvollen Quellen
Paper from responsible sources
FSC® C105338
www.fsc.org

If you have any concerns about our products,
you can contact us on
ProductSafety@springernature.com

In case Publisher is established outside the EU,
the EU authorized representative is:
Springer Nature Customer Service Center GmbH
Europaplatz 3, 69115 Heidelberg, Germany

Printed by Libri Plureos GmbH
in Hamburg, Germany